FUNDAMENTAL PARTICLES

an introduction to quarks and leptons

The Super Proton Synchrotron at CERN, Geneva.

FUNDAMENTAL PARTICLES

an introduction to quarks and leptons

B. G. Duff

Taylor & Francis
London and Philadelphia
1986

UK	Taylor & Francis Ltd, 4 John St, London WC1N 2ET
USA	Taylor & Francis Inc., 242 Cherry St, Philadelphia, PA 19106–1906

British Library Cataloguing in Publication Data

Duff, B. G.
    Fundamental particles: an introduction to
    quarks and leptons.
    1. Particles (Nuclear physics)
    I. Title
    539.7′21   QC793.2
    ISBN 0–85066–332–6

Phototypesetting by Georgia Ltd, Liverpool
Printed in Great Britain by Taylor & Francis (Printers) Ltd,
Basingstoke, Hants

FOREWORD

It is hoped that this book will be useful to undergraduates as a first introductory reader in the subject of particle physics before they go on to a formal course (usually in the third year) when a more advanced text (such as *Introduction to High Energy Physics* by D. H. Perkins, Addison Wesley, second edition, 1982) will be required. It may also be of interest to advanced sixth-formers or older scientists who wish to be kept informed of current developments in fundamental particle physics. Chapters 1 to 4 and 7 may be of particular interest to such readers, as they give a more or less self-contained elementary account of the fundamental particles. Chapters 5 and 6 go into rather more of the technical details concerning the electroweak and strong interactions and are a little more advanced than the rest of the book. As far as possible, the mathematical details have been kept to a minimum. It is also assumed that the reader has only a limited acquaintance with quantum mechanical ideas and special relativity. Summaries of these topics are given in Appendices at the end of the book for the interested reader.

CONTENTS

PREFACE

Particle physics has changed dramatically in the last decade or so, and in many ways the subject is now much more ordered. This book attempts to give a simple introduction to the subject using the current view of particle physics which is embodied in what is often called the 'Standard Model'. Although we are very far from any complete understanding, many aspects of the nature of fundamental particles are much better understood, as was exemplified in the historic discovery in 1983 of the W and Z particles whose existence and properties had been predicted within the framework of the standard model. As I write this preface, a preliminary indication of the discovery of the 'top' quark has been announced at CERN, suggesting that another important piece in the jigsaw puzzle of fundamental particles has been found.

Very few historical details are given as this no longer seems to be a useful approach, at least when first encountering the subject. Many of the early theories of fundamental particles have now been overtaken or included in the new models, and attempting to follow the tortuous routes that have led to the current understanding of particle physics can be unrewarding, at least for a newcomer to the subject!

The study of fundamental particles explores not only the subnuclear structures of matter but also the laws which govern the forces as they operate at high energies and correspondingly small distances. The newly gained knowledge of these forces is probably even more important than the developments that have led to understanding at a deeper level the structure of matter. One of the attractions of this subject is that studying the basic constituents can give rise to new laws of nature as well as models for sub-nuclear structure.

The book emphasizes the basic physical models of elementary particles in terms of the quarks and leptons and discusses in a simple way the forces which act between them. By and large, very few experimental details are given as these can cloud the descriptions of the significance of the results obtained. Modern high-energy physics experiments involve equipment and laboratories on the scale of large factories and can be very intimidating. However, it is not necessary to understand the full technology of accelerators and particle detectors to appreciate the remarkable developments which have occurred recently in the understanding of the structure of matter in its most basic forms, and the forces which operate between the constituents.

To Marjorie

Oliver and Harriet

ACKNOWLEDGEMENTS

I would like to thank many of my colleagues at University College London and CERN for discussions which have helped me to write this little book. In particular, I am indebted to Dr. Donald Davis and Dr. Colin Wilkin who read the manuscript and made a large number of useful suggestions and comments. I would also like to thank Una Campbell for the preparation of the line drawings.

I am grateful for permission to use photographs supplied by CERN (frontispiece, Figures 1.2(*a*), 1.3(*a*), 1.4, 2.1, 5.6 and 7.3), by DESY (Figures 1.2(*b*) and 6.15), by SLAC (Stanford Linear Accelerator Center, run by Stanford University for the US Department of Energy) (Figure 1.2(*c*)), by Dr. Frederick Bullock of University College London and the Gargamelle collaboration (Figures 1.3(*b*) and 5.4) by Dr. Donald Davis of University College London and the WA17 Collaboration (Figure 4.6) and by Professor Peter Kalmus of Queen Mary College and the UA1 collaboration (Figure 5.7).

University College London   Brian G. Duff
July 1984

Sadly, Brian died suddenly in the final stages of preparation of this manuscript. We have made only slight corrections and amendments for publication and in this we have received useful advice from Dr. David Bailin. We acknowledge with thanks the technical assistance of Mrs. Teresa Debono.

University College London   D. H. Davis
November 1984   C. Wilkin

CHAPTER 1

fundamental ideas

1.1. The domain of particle physics

The physical phenomena that can be studied today span an enormous range of space and time. The shortest distances that can be penetrated are about 10^{-18} m but the furthest known galaxies are 10^{26} m away. Similarly, the shortest-lived particles have a lifetime of about 10^{-24} s while current estimates of the age of the Universe are of the order of 10^{17} s. We thus have over 40 orders of magnitude of space and time in which to study the physical laws of nature.

Everyday phenomena are dominated by the time and space scales of the human race itself. Times with which we are most familiar range from our minimum bodily response times (of the order of 10^{-3} s) to our lifetimes (of the order of 10^9 s). The resolution of our optical microscopes is around 10^{-6} m and the circumference of the Earth is about 10^8 m. Thus, our normal perceptions span only about one quarter of the 'available' orders of magnitude in space and time. We should not be too surprised if this classical physics domain does not give a complete picture of the totality of physical phenomena. In particular we would expect that phenomena on the edges of the known space–time domain would not look too familiar to us. Hopefully, these are also the most likely regions to be studied in order to discover new basic laws of nature.

In that sense, astronomy and particle physics are usually regarded as among the most fundamental areas worthy of pure research study. The astronomers and space scientists explore the large-distance, large-time domains and the particle physicists explore the small-distance, small-time domains. Astronomical and cosmological studies require an understanding of the long-time development of massive systems and effects from general relativity become important. On the other hand, the world of sub-nuclear particles is dominated by quantum mechanical effects and special relativity.

We do not expect, of course, to throw away our basic classical ideas but rather to see them as a special limiting case of some general theories. For example, the classical relativity that we are familiar with in everyday life is encompassed within special relativity and is the limiting case when velocities are much less than the speed of light in a vacuum. Similarly, Newtonian mechanics is a limiting case of the more general quantum mechanics which is an essential feature in the understanding of small-scale phenomena.

Many of the most important laws of the classical domain are carried over into the small-scale regions. For example, the fundamental conservation laws of linear and angular momentum and energy are maintained, as are our basic ideas about electromagnetism.

1.2. The structure of matter

In everyday life, we see all about us material existing in solid, liquid and gaseous forms. Ice, water and water vapour are all familiar forms of large assemblies of molecules each of which has the same chemical formula, H_2O. The possible arrangements of the molecules together with the conditions, such as temperature and pressure, which lead to a particular stable phase of matter are not the subject of this book, but form an important part of basic physical studies. Here we are concerned with the constituents themselves, rather than the large assemblies which must be treated in a statistical manner.

The water molecule itself consists of three atoms: one oxygen atom and two hydrogen atoms. They are held together by chemical bonds which originate from the residual electromagnetic forces between the atoms.

In its turn, an atom consists of a positively charged heavy nucleus surrounded by negatively charged electrons. The stability of such an atom again depends upon its constituents being in a bound state allowed by the electromagnetic forces between the nucleus and the electrons. The obvious fact that most material is stable is one of the most important reasons for believing that the classical laws of physics are not sufficient for the understanding of atomic structures. If we take a simple classical view of a hydrogen atom as a negatively charged electron circling round the positively charged nucleus (just a proton in this example), then because the electron is in an orbit it must be accelerating all the time. All accelerated charges radiate and therefore the electron must be continuously losing energy. A quick calculation shows that classically the electron spirals quickly in towards the proton which attracts it, and all such atoms would vanish, emitting a flash of light, after about 10^{-10} s! If all atoms are so unstable, how are we all still here?

The answer to the stability problem lies in quantum mechanics, where it can be shown that a system such as the hydrogen atom can have many allowed bound states with different stability and the lowest energy state allowed is the most stable system. The vast bulk of atoms in normal material exist in their 'ground states' which are stable.

The heavy nucleus of any atom is made up of constituent particles called nucleons, which are of two types. First, there are the protons which are positively charged and, second, there are neutrons which are electrically neutral and have about the same mass as the protons.

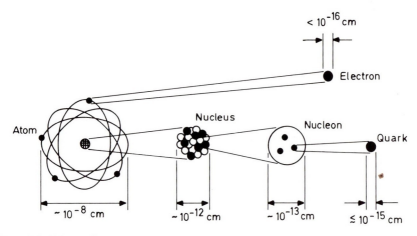

Figure 1.1. Sub-atomic structure.

In the last decade it has become increasingly clear that the proton and neutron are by no means the ultimate building blocks of matter but in their turn are composed of assemblies of objects of nearly point-like dimensions called 'quarks', bound together through the 'strong' nuclear force. (These 'quarks' were named by Murray Gell-Mann from an enigmatic quotation from James Joyce's *Finnegan's Wake*: "Three quarks for Muster Mark!") The subject of fundamental particles has now come to mean the study of the properties of the smallest known constituents of matter, such as quarks and electrons, and the forces which govern their interactions.

Table 1.1. Sizes of the constituents of matter.

		Typical dimensions
MOLECULES		10^{-7}–10^{-8} cm
ATOMS		about 10^{-8} cm
NUCLEI		about 10^{-12} cm
NUCLEONS		about 10^{-13} cm
QUARKS	ELECTRONS	$< 10^{-15}$–10^{-16} cm

At the present time there is no evidence for any internal structure within quarks or electrons and for our present purposes we shall assume that these particles can be regarded as point-like fundamental constituents. Notice that the volumes occupied by atoms, and even nucleons, mostly consist of just empty space!

1.3. How can sub-nuclear particles be studied?

Although this book will not attempt to give many experimental details as to how we can study fundamental particles, it should be realized that the whole subject is based firmly on a vast amount of experimental data accumulated over the last few decades. The theoretical ideas, elegant though they may be, would be of little interest unless they properly describe the real physical observations that have been made.

If we try to study the structure of matter using a microscope, we will clearly be limited by the wavelength of the light used. One cannot hope to see any structures of dimensions less than one wavelength of the probing light. Thus, the shorter the wavelength, and correspondingly the higher the frequency, the better from this point of view. High frequency implies high energy and we immediately see the connection between short distances and high energies. The biologist uses an electron microscope employing relatively high-energy electrons as the probes for structures which can be as small as one large molecule. The corresponding de Broglie wavelength ($\lambda = h/p$, see Appendix B.1) will be short enough to study such small structures. If we wish to study sub-nuclear, or even sub-nucleonic structures, exceedingly high-energy probes (usually electrons or protons) will be required. The particle physicists' 'microscopes' are huge accelerators capable of producing extremely high-energy particles. (For this reason particle physics is often called 'high-energy physics'.) With the highest accelerator energies available today, distances of the order of one thousandth of the proton's radius have now been resolved.

Modern accelerators are so large, and therefore expensive, that a limited number of laboratories have been set up in various parts of the world. These are mainly international centres where high-energy physicists from many countries are welcomed.

Within Western Europe, the main accelerator centres are based in Geneva (the CERN laboratory) and in Hamburg (the DESY laboratory). There are a number of accelerator laboratories in the USA, including the Stanford Linear Accelerator Center in California, and the Fermi National Accelerator Laboratory near Chicago. Other centres exist in the USSR and Japan. In all cases, large collaborations of physicists share these huge facilities. A wide range of highly sophisticated accelerators are now available, offering various particle types and energies as probes of matter (see, for example, Figure 1.2). The energies of the particles from the various machines are usually quoted in gigaelectronvolts (GeV) (see Appendix A.2 for a note on units).

Another important reason for going to high energies, other than the need for short-distance probes, is that they are also required for the production of interesting new particles. Classically, we talk about 'elastic' collisions, such as those between billiard balls, and 'inelastic' collisions, such as those between two cars that collide head-on. In the car crash a great deal of energy is given up in heat and a little in sound and a large number of small pieces of cars litter the road!

Figure 1.2(*a*). 400 GeV Super Proton Synchrotron (SPS) at CERN.

Figure 1.2(b). The DESY laboratory (Hamburg). The 'PETRA' electron-positron collider gives an available energy of more than 40 GeV. The outline of the new 'HERA' machine is shown by the broken line.

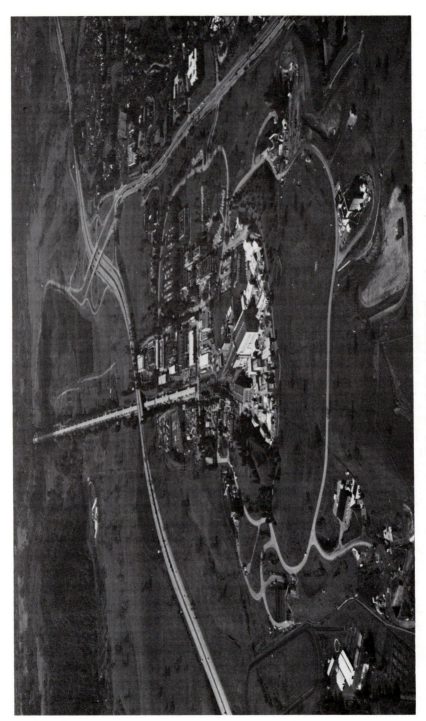

Figure 1.2(c). Stanford Linear Accelerator Center (near Palo Alto, California) (20 GeV linear electron accelerator can be seen).

Figure 1.3(*a*). The 3·7 m Big European Bubble Chamber (BEBC) (usually filled with liquid hydrogen).

Classically, such a collision is not very interesting (unless you happen to be personally involved). However, the same kind of collision at the particle level results not in broken pieces of particles, but in the production of brand new and often more complex particles. It is as if our road crash resulted in a collection of Mercedes, Rolls Royces and the occasional London bus! We will go into this particle production process further when we have discussed the conversion of energy into mass.

Many of the most interesting aspects of particle physics are the rules which govern such production processes. We must understand which properties must be conserved and what new features can appear in such collisons. In general, all our knowledge of sub-nuclear particles and the laws which govern their behaviour comes from experiments in which we scatter one particle off another. The accelerators provide the incident particles. Targets may be a wide range of objects. For example, we may choose liquid hydrogen, which has the advantage of reasonably high density and very simple atoms consisting of just single protons and electrons. The atomic electrons can induce only very small-angle scattering since they are far less massive than the proton. On the other hand, if the process we are interested in is very rare, we may choose a very dense target like tungsten to

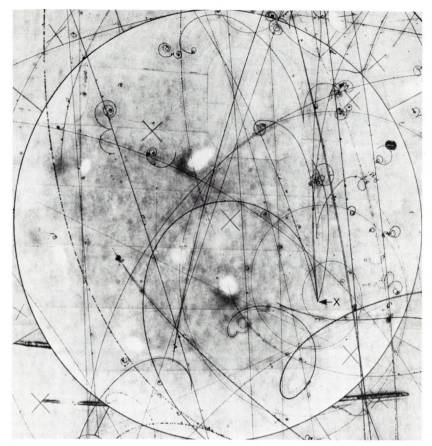

Figure 1.3(*b*). A bubble chamber photograph taken in BEBC showing a neutrino interaction (X = interaction point).

provide a large collision rate, but then we would have to put up with the complications of understanding scattering off complex nuclei.

In order to study the outgoing particles we require detectors for the energetic and often charged products. Very frequently we make use of the fact that such energetic charged particles lose a very small proportion of their energy as they pass through matter by knocking out some of the molecular electrons, leaving positively charged ions behind. The presence of this separation of electric charge in the wake of the fast particle is known as ionization, and is at the root of most particle detection systems.

Two general groups of detectors are commonly used. Those in the first group are called 'visual' detectors and the best known device in this category is the bubble chamber (see Figure 1.3(*a*)). Liquid is maintained just below the boiling condition and a rapid pressure drop is produced by a piston causing a sudden expansion, just before a beam of particles passes through. Under these conditions the liquid becomes superheated and a trail of bubbles will form along the path of ionization caused by the particle beam. This track of closely spaced small bubbles can then be photographed. Collisions will sometimes occur within the liquid of the detector itself, so that the tracks from secondary particles will also be seen. Very detailed information about such an event can then be obtained by measuring the photographs. Several views are usually taken so that three-dimensional reconstruction can be achieved and usually the whole chamber is within a region of high magnetic field so that momenta of the charged particles can be extracted by measuring the radii of curvature of the tracks (see Figure 1.3(*b*)).

Devices in the second group are called 'electronic' detectors. In some detectors of this kind, the deposited charge is collected and amplified by localized electrodes which then give the position of the particle. In some materials the charge can produce scintillations of light which can be picked up by light guides, converted into an electrical signal by a photo-cathode and the electrical signal then amplified. Another process, known as the Čerenkov effect, produces a cone of blue light as a shock wave when a particle traverses a medium at a speed faster than that of light in the medium. This is exactly analogous to the sound shock wave produced by an aeroplane as it passes through the sound barrier. The cone angle of the Čerenkov light enables the velocity of the particle to be calculated.

The momenta of charged particles can be found by tracking them through magnetic field regions. Thus, combinations of magnetic spectrometer tracking systems and Čerenkov detectors can give the masses of the outgoing particles to help to identify them. Many other sophisticated particle detectors have been devised and the interested reader is directed to the list of further reading at the end of the book.

1.4. The forces

Matter in various forms experiences four basic forces. The interactions and resulting motions of very large objects such as the Sun, Earth and Moon are essentially entirely governed by the gravitational force. The gravitational attraction of the large mass of the Earth to the small masses clinging to its surface, such as men, are obviously of dominant importance in everyday life on this planet. However, paradoxically, gravity appears to play almost no part at all in the interactions between the smallest constituents of matter. The strength of this force is so small compared to the other known forces (see Table 1.2) that it plays an insignificant role in the detailed behaviour of the quarks and electrons. In fact, so far no experiment has been able to detect any effects arising from gravitational interactions between fundamental particles.

The electromagnetic force is responsible for the atomic and molecular structure of matter and therefore ultimately governs the whole areas of chemical and biological structures, including man himself. Moreover, we shall see later that many of the building blocks of matter carry electric charges and there are

Figure 1.4. The huge electronic detector system, UA1 at CERN (see chapter 5).

very important electromagnetic effects at the sub-atomic and sub-nuclear levels.

The strong nuclear force holds the quarks within the nucleons and is also responsible for binding together the nuclear constituents, allowing nuclei to be stable despite the positive charges present within the assembly of nucleons. It must be a strong force to overcome the electromagnetic repulsive forces which we know will occur. It is this force which is seen in its most fundamental form in the interactions between the quarks.

Finally, the weak nuclear force is required to give an understanding of a class of phenomena typified by the beta radioactive decays of certain nuclei. This force is also responsible for the initiation of the process which leads to the steady release of the Sun's energy.

Each of these four forces has a dominant effect in certain regimes and is characterized by very different relative strengths for processes governed by the particular force (see Table 1.2).

The distances over which the forces are significant are also indicated in Table 1.2. The gravitational and electromagnetic forces vary as the inverse square of the distance away from the source and therefore fall off relatively slowly (we say that such forces have an 'infinite' range). The weak and strong forces are only significant over very small distances and further out they have a negligible effect (so they are 'short-range' forces).

The relative strengths of the four forces given in Table 1.2 are a measure of their typical values. As their ranges are very different, each of the forces dominates in various spatial domains. The strong force is overpowering within the nucleon, but at 10^{-8} cm the coulomb force has taken over as the dominant contribution. However, once electrically neutral atoms are centimetres apart, the only significant force left is gravitation.

Table 1.2. Fundamental forces.

Force	Relative strength	Range	Important for
Gravitation	10^{-38}	Infinite	Heavenly bodies
Weak nuclear	10^{-5}	10^{-16} cm	Radioactive decay
Electromagnetic	10^{-2}	Infinite	Atomic structures (residual effects in molecular binding)
Strong nuclear	1	10^{-13} cm	Quarks within nucleons (residual effects in nuclear binding)

Strong attractive force
binds the nucleus

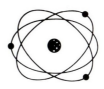

Electromagnetic force binds
negatively charged electrons
to positively charged nucleus

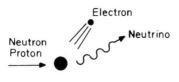

Weak force:
Radioactive decay

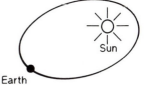

Gravitational force: matter
attracted. Solar system

Figure 1.5. The fundamental forces.

One of the great questions of modern physics relates to the need for four separate forces. Just as Maxwell successfully combined the laws of electricity and magnetism into one common set of equations, so we would hope for the economy of a single description for all the basic forces of nature. Recently, the first step in this direction has been made and a comparable achievement to that of Maxwell has been obtained by Steven Weinberg, Abdus Salam and Sheldon Glashow, who have put electromagnetism and the weak force into a common framework (see chapter 5). Many physicists are trying to extend these ideas to take in the strong force and even gravitation and we shall return to this problem in the final chapter.

1.5. Which particles 'feel' which forces?

Although gravitation was the first interaction to be studied systematically and recognized as a basic force, surprisingly it is the least understood at a fundamental level. What is clear, as mentioned before, is that it is too feeble to be detected in particle interactions on an individual basis. However, at the macroscopic level the mass of an object gives it the possibility of a gravitational interaction with another object also having mass[†]. Of course, this mass is derived from the masses of the individual constituents. We believe that the gravitational force always acts attractively.

In a similar way, all particles carrying a net charge (or at least a distribution of charge) will feel electromagnetic forces, which may be attractive or repulsive depending on the signs of the charges. Even though two atoms may each be electrically neutral when viewed from sufficiently far away, when atoms become close to each other the forces arising from the electrons and nuclear charges may allow the formation of stable molecules.

One large family of particles feels the weak force, but not the strong nuclear force. Such particles are referred to as leptons. An electron is an example of a

Table 1.3. Particles and forces.

Particle	Strong force	Electromagnetic force	Weak force	Gravitational force	Class
Proton	Yes	Yes	Yes	Yes	Hadron
Neutron	Yes	Yes	Yes	Yes	Hadron
Electron	No	Yes	Yes	Yes	Lepton
Neutrino	No	No	Yes	Yes[a]	Lepton
Quark	Yes	Yes	Yes	Yes	Hadron

[a]*see footnote.*

---

† A particle with zero 'rest mass' (see section 2.3), such as the photon, with energy E, will have a momentum E/c and a relativistic mass E/c^2 (see Appendix A.3) and hence this energy is associated with inertia. Thus 'massless' particles such as the photon are influenced by a strong gravitational force through the coupling to its energy and momentum.

particle in this class. (Notice that the electron also feels gravitation, albeit extremely feebly, and as it carries a negative charge it also certainly experiences electromagnetic interactions.) One interesting group of leptons are called neutrinos. These carry no electric charge and therefore cannot feel any electromagnetic forces and as they are leptons they do not experience the strong force either. This means that in collisions with quarks, for example, they will interact through the weak force. Such interactions will be felt only over a very short range.

All those particles which experience the strong force are collectively called the hadrons. Protons and neutrons are familiar examples of this group. At the most fundamental level, the quarks are also hadrons.

CHAPTER 2

basic particle properties

2.1. The quantum nature of fundamental particles

The ideas of quantum mechanics tell us that it is not possible to think of very small particles, whether they are molecules, atoms, nuclei or quarks, just as well defined objects, rather like minute billiard balls. There is a wealth of experimental evidence that shows that all particles have a wave-like nature. Conversely, it is also clear that waves have a particle aspect. This 'wave–particle dualism' is of fundamental importance in obtaining a picture of the elementary particles.

It is assumed that electromagnetic radiation of frequency v is equivalent to a stream of particles (photons) each having an energy E given by

$$E = hv \qquad (2.1.1)$$

where h is Planck's constant. As the photons are thought to be massless it then follows (see Appendix B.1) that each photon carries a momentum p given by the relation

$$p = h/\lambda \qquad (2.1.2)$$

where again h is Planck's constant and λ is the wavelength. It follows that light treated as a stream of photons can impart momentum to an object it strikes. This idea is central to the understanding of a wide range of phenomena in physics such as the photoelectric effect, the Compton effect, black-body radiation and the discrete nature of the energy changes that give rise to atomic spectra.

Conversely, when electrons are scattered off a crystal lattice, diffraction effects are observed. Phenomena like interference and diffraction are an immediate indication of a wave-like nature and could never be understood in terms of a purely particle-like behaviour. Such phenomena can be quantitatively understood if it is assumed that a particle having a momentum p is associated with a wave with wavelength λ given by

$$\lambda = h/p \qquad (2.1.3)$$

(cf. equation 2.1.2)

In this picture, a single particle can be seen as a wave-packet, that is, as a wave which dies away at large distances or times. One is then immediately faced with the problem of the significance of the amplitude of the wave, which we shall call Ψ. For ordinary waves such as sound the intensity observed is proportional to the square of the amplitude. However, when there are two sources of sound it is the amplitudes which are additive, and their relative phase gives rise to interference phenomena such as beats. This superposition principle leads to analogous effects in quantum mechanics. In quantum mechanics it is assumed that the amplitude squared gives a measure of the mean particle density. Thus, if the 'wave-function' (amplitude of the wave) is Ψ, we say that the probability, P, of finding the electron in an elemental volume dv is given by

$$P = |\Psi|^2 \, dv \tag{2.1.4}$$

The wave-function thus tells us where the electron is likely to be found and with what probability. It is therefore sometimes referred to as a probability wave or a matter wave. (There is a slight complication as compared to sound waves since Ψ may be a complex number. In equation 2.1.4 we have to take the modulus squared of Ψ, that is, the sum of the squared real and imaginary parts — see Appendix B.)

On the other hand, an interaction with another object will depend upon the properties carried by the particle such as its electric charge. The unit of electric charge cannot be broken up and so in the actual interaction it is more useful to think of the particle-like nature of the object.

In classical physics we are used to the idea that waves can sometimes be described by discrete states. For example, a finite length of string, held at its two ends, can vibrate in its fundamental mode or any of its harmonics. A single integral number is sufficient to define a particular allowed mode of vibration. It can also vibrate in a mixture of these allowed modes of vibration. In this situation it is clear that only certain frequencies are allowed, not a continuum of frequencies. Two-dimensional bounded systems such as a rectangular plate, or three-dimensional objects such as a block of jelly-like material, also vibrate at definite frequencies determined by the size and properties of the object. In such a two- or three-dimensional case, it is found that two or three integers, respectively, are required to define uniquely a mode of vibration. (In these cases it is also found that not all combinations of integers lead to a unique frequency of vibration, and hence energy, of the system. When several modes of vibration have the same frequency, we say there is degeneracy.) Of course, if boundaries are not well defined, such as in the case of a travelling plane wave, then all frequencies are allowed and there is a continuum of possibilities.

One of the triumphs of quantum mechanics is the explanation of the existence of discrete states in 'particles'. These quantum states are particularly striking in the case of atoms. The spectral lines observed as definite colours (that is, frequencies) must correspond to the emission of light when an atom falls from one high energy quantum state to a lower one. The frequency, ν, of the emitted light will be given by

$$\nu = (E_2 - E_1)/h \tag{2.1.5}$$

where E_2 and E_1 are the higher and lower energies, respectively. It follows that a hydrogen atom, for example, exists in definite quantum states. The electrons of the atom are bound to the nucleus through the coulomb potential (electromagnetic attraction) and are, in a sense, so confined in a rather complicated three-dimensional box. By analogy with a classical wave in such a box, we would expect discrete states to be allowed (see Appendix B.2). Moreover, we would expect there to be three numbers required to define the orbital state of the electron. This indeed turns out to be the case and three integers (usually called *n, l* and *m*) are needed to define the quantum orbital state of the hydrogen atom. These numbers are referred to as quantum numbers and we will come across many other examples of such quantities in our considerations of the fundamental particles.

2.2. The stability of particles

Some of the most familiar particles, such as the proton and the electron, seem to be completely stable. It is clear that most material in everyday life does not spontaneously decay! However, even the proton is known to be stable only within certain rather large limits (see chapter 7). In many cases, the more stable particles are thought of as 'ground-states' of whole families of particles with similar properties. They tend to be the particles with the lowest mass in the family.

However, many particles with higher mass can decay due to the influence of one or more of the strong, electromagnetic or weak forces. Decays through the strong force will occur in extremely short times (typically of the order of 10^{-23} s), whereas electromagnetic and weak decays occur much more slowly, typically in the range 10^{-20} s to 10^{-6} s (see Table 2.1).

If the probability per unit time of a particle decaying is λ and there are N particles of which dN decay in a time dt, then

$$dN = -N\lambda\,dt$$

and integrating

$$N = N_0 e^{-\lambda t}$$

where N_0 is the initial number of particles. The mean lifetime of the particle, τ, is defined by

$$\tau = 1/\lambda \tag{2.2.1}$$

and so we have

$$N = N_0 e^{-t/\tau} \tag{2.2.2}$$

It can then be seen that τ is the time at which the number of particles has been reduced to $1/e$ of the initial number.

According to Heisenberg's Uncertainty Principle (see Appendix B.1) it is not possible to know exactly the energy of any particle at a precise instant. One consequence of this is that an unstable particle decaying in a time Δt is associated with an uncertainty in energy $\Delta E \sim \hbar/\Delta t$, where $\hbar$ is Planck's constant h divided by 2π (approximately $6\cdot6 \times 10^{-22}$ MeV s). Thus the energy given up in the decay is not precise and there will always be a distribution of energies in the system of particles resulting from the decay. Hence, a particular lifetime is associated with a characteristic 'width' Γ in the energy spectrum for the decay process. For example, a strong decay resulting in a lifetime of $\sim 10^{-23}$ s is associated with a broad energy width of the order of several hundreds of MeV. On the other hand, a charged pion has a width of only about 3×10^{-8} eV, which means that its mass is uncertain to 3×10^{-8} eV/c^2.

Table 2.1. Examples of particle lifetimes and corresponding widths of the energy uncertainties for the decay.

Particle	Approximate lifetime (seconds)	Width (MeV)
Electron	Stable	
Proton	Stable	
Neutron	898	
Muon	$2\cdot2 \times 10^{-6}$	3×10^{-16}
Charged pions	$2\cdot6 \times 10^{-8}$	$2\cdot5 \times 10^{-14}$
Charged kaons	$1\cdot2 \times 10^{-8}$	$5\cdot5 \times 10^{-14}$
$D^{\pm}$, D^0 (charmed)	10^{-13}–10^{-12}	10^{-8}–10^{-9}
Neutral pion	$0\cdot8 \times 10^{-16}$	8×10^{-6}
Rho meson	4×10^{-24}	160
Δ (1236) baryon resonance	6×10^{-24}	115
N* (1470) baryon resonance	$2\cdot2 \times 10^{-24}$	300

2.3. Particle mass

Classically, mass is a measure of inertia, that is, how difficult it is to get an object to change its motion. But it is also a measure of the strength of the force felt by the object in a gravitational field. The units are chosen so that both definitions give the same numerical value for the mass of an object.

Within the framework of special relativity, we say that the mass of an object depends on its velocity, v, so that its relativistic mass, m, is given by

$$m = m_0/(1 - v^2/c^2)^{\frac{1}{2}} \tag{2.3.1}$$

where m_0 is the 'rest mass' — the mass when the speed is zero. From this it is clear that if v is much less than c, the velocity of light in vacuum, the relativistic mass tends to the rest mass, m. Thus, in the classical limit we use only a rest mass.

Most particles, like the electron, proton and neutron, have a mass when they are at rest. However, the photon always travels with the velocity of light and it never can be at rest. Special relativity (see Appendix A.3) tells us that there is an equivalence between mass and energy and therefore a photon can acquire mass through its motion. In general we can write an equation connecting the energy, E, of a moving particle with its momentum, p, and its rest mass, m_0, as follows:

$$E^2 = p^2 c^2 + m_0{}^2 c^4 \qquad (2.3.2)$$

(or $E = mc^2$, where m is the relativistic mass; see Appendix A.3). In the special case of a particle at rest, this reduces to the famous relation

$$E = m_0 c^2 \qquad (2.3.3)$$

giving the equivalence between rest mass and energy.

The electron is the lightest particle with a well-measured rest mass and it is often convenient to use this mass as a basic unit. Thus, for example, the proton mass can be expressed as 1836·1515 electron masses. The relation 2.3.3 can also be used to express the proton mass in terms of equivalent energy units. The proton rest mass is equivalent to an energy of 938·280 MeV (see Appendix A.2).

A number of fundamental particles, notably the photon and the various types of neutrino (see section 4.7) have a negligible (and probably zero) rest mass. Such particles travel at the speed of light and acquire momentum by virtue of their motion. For such massless particles, equation 2.3.2 combined with equation 2.1.1 gives

$$E = pc = h\nu$$

and so
$$p = h\nu/c = h/\lambda \qquad \text{(already seen above as 2.1.2)}$$

and the relativistic mass will be given by

$$m = E/c^2 = h\nu/c^2 \qquad (2.3.4)$$

Attempts have been made to determine the upper limit for the mass of neutrinos. For example, a careful study of the tritium beta-decay spectrum could reveal small effects that would arise from the (electronic) neutrino having a finite mass. In some experiments masses of up to 35 eV/c^2 have been claimed, but these results are not yet fully confirmed and for the moment we shall assume that the neutrinos have zero mass. (Incidentally, there are very interesting cosmological consequences of finite-mass neutrinos. Such 'massive' neutrinos would seriously affect the gravitational potential, and hence energy, in a galaxy and cosmological models would need to take such effects into account. In some circumstances, the existence of neutrinos with mass could lead to oscillatory behaviour in the Universe instead of continuous expansion.)

Many of the composite particles we are interested in, and even some of the basic constituents, have large rest masses, and in order to produce such objects and study them correspondingly large energies are required. (We have already mentioned that in order to study very small objects large energies are required, but now we have this second reason why particle physicists are always trying to push up the energies available to them in accelerators.)

In some accelerators, two beams of particles are stored and allowed to collide more or less head-on (see Figure 2.1). In such 'intersecting storage rings' the available energy for the production of new particles is simply the sum of the two individual total energies, as the accelerator is essentially in the centre-of-mass (cms) coordinate frame. In a 'fixed-target' system an energetic beam of particles, with energy E, hits a stationary target made up of particles of mass m, and the available energy, $\sqrt{s}$, is

$$\sqrt{s} \sim (2\,mE)^{\frac{1}{2}} c \qquad (2.3.5)$$

as shown in Appendix A.3. Whilst the fixed target machine obviously operates with a much smaller available energy, it does have the advantage of the possibility of a very dense target, whereas the intersecting storage rings have two very low-density beams and the collision rate is therefore much lower.

2.4. Spin and angular momentum

In general, a particle can have a 'bulk' motion with some linear momentum, $\mathbf{p}$, and possibly some orbital angular momentum, $\mathbf{L}$. In addition, particles may be spinning with a resulting spin angular momentum, $\mathbf{S}$. The total angular momentum, $\mathbf{J}$, is given by

$$\mathbf{J} = \mathbf{L} + \mathbf{S} \qquad (2.4.1)$$

A classical particle can take any values for these various momenta, but quantum mechanically a particle is restricted to definite discrete angular momenta. The allowed spin angular momenta are defined by the following results from quantum mechanics.

The magnitude of the spin angular momentum is

$$|\mathbf{S}| = [s(s+1)]^{\frac{1}{2}}\,\hbar \qquad (2.4.2)$$

where s is the quantum number specifying the spin (not to be confused with the term $\sqrt{s}$, traditionally used for energy, in equation 2.3.5).

Figure 2.1. The Intersecting Storage Rings (ISR) at CERN. (Protons collide nearly head-on at the crossing point of the vacuum tubes.)

We can also measure the component of the spin along some direction, which is usually taken to be the z or third axis.

$$S_z = m_s \hbar \qquad (2.4.3)$$

where s is integer or half-integer and $m_s = -s, -(s-1), \dots s$.

Thus, for an electron with $s = 1/2$, the z-projection can only take values $+$ or $-1/2$ (in natural units, see Appendix A.2). We often speak of 'spin-up' and 'spin-down' electrons, but this is rather loose and a more correct picture is given in Appendix B.4. On the other hand, a massive particle with $s = 1$ can have three z-components: -1, 0 or $+1$. In this case there are three orientations for the spin vector.

If we consider a small finite spinning charge from a classical point of view, the small current loop would generate a magnetic field. The particle behaves as if it had a magnetic dipole moment μ, such that

$$\mu = \frac{ge\hbar}{2m} \left(\frac{\mathbf{S}}{\hbar} \right) \qquad (2.4.4)$$

where g is a numerical factor, usually called the g-factor (or Landé factor). The unit is the magneton, defined by $(e\hbar/2m)$ where m is the mass of the electron for the Bohr magneton and the proton mass for the nuclear magneton; e is the unit electric charge on the electron or proton.

The spin quantum number, s, divides all particles into two main groups. Particles with half-integral spin quantum number s, such as the electron and the proton, are known as fermions, whereas particles with integral (including zero) values of s are called bosons.

It seems that, at the most fundamental level, particles which are exchanged in order to carry a force are all bosons, but particles which are interacting through such exchanges are themselves fermions. Thus the quarks are fermions and the gluons, the particles which mediate the strong force between quarks, are bosons. Similarly, the recently discovered massive exchange particles ($W^{\pm}$ and Z^0) in the weak interaction are referred to as intermediate vector bosons, as explained in section 3.1.

A special quantum mechanical rule applies to fermions. This rule is known as the Pauli exclusion principle and states that no two identical fermions in an atom can be in exactly the same quantum state (that is, they must have at least one different quantum number). This idea provides the basis of our understanding of the way that the allowed ground states of atomic systems can be built up as more electrons are brought into atomic structures. Hence, it is central to our ideas of the origin of the periodic table of elements and the resulting chemical properties. This rule docs not apply to bosons, and as many bosons as you like can all exist with precisely the same set of quantum numbers. (The fundamental idea of fermions and bosons is related to the symmetry of the combined wave-function for two identical particles under the exchange of the particles. This concept is briefly discussed in Appendix B.5.)

2.5. *Electric charge and isospin*

It had been realized in the 1930s that the protons and neutrons which are the constituents of the nucleus have very similar masses and it is natural to think that perhaps the proton and neutron are simply two different charge states of the same basic particle, often called collectively the nucleon. These two charge states were seen as an analogue of the two possible spin states (orientations) of a spin 1/2 fermion (see Figure B.1(*a*), p. 115). By analogy, we talk about an 'isospin' quantum number, $I = 1/2$ with isospin up ($I_z = +1/2$) or isospin down ($I_z = -1/2$). The charges of the two states of the nucleon are related to I_z by

$$Q = 1/2 + I_z \quad (= 0 \text{ or } 1) \tag{2.5.1}$$

It seems that the strong interactions do not distinguish between these two states, unlike the electromagnetic interactions.

Another way of saying this is to remark that the strong interaction conserves isospin, that is, it depends upon I but not upon I_z. Thus the strong force does not know whether it is dealing with a proton or a neutron. The electromagnetic interaction 'sees' the third component, I_z, and obviously its effect depends upon the charge states involved, but does not conserve isospin.

Some common particles (such as the pi-meson or pion) are found to exist in three charged states (-1, 0 or $+1$ on the charge of the electron) and therefore are assigned isospin $I = 1$ by analogy with spin $s = 1$ particles which have three spin orientations (see Figure B.1(*b*)). In these cases equation 2.5.1 for the charge must be generalized to

$$Q = B/2 + I_z \tag{2.5.2}$$

where B is called the 'baryon number' and is defined as $B = 1$ for nucleons and $B = 0$ for mesons which feel the strong interaction, such as pions (see section 4.2).

2.6. Conserved quantities

In classical physics we are familiar with a number of conservation laws. In mechanics we meet the conservation of energy, linear momentum and angular momentum. These most general laws are also of great importance in fundamental particles physics and we understand their origins in terms of very basic properties of space and time. We are also used to the idea of electric charge being conserved and again this law seems to be universal.

We believe that space is homogeneous (uniform) in the sense that there is no 'structure' in some empty region of space which would somehow change the laws of physics in that region. Of course, we would have to assume that the region is field-free as obviously objects will be affected differently by the presence of various gravitational, electric, magnetic or other fields depending upon the properties of the objects. This means that objects that undergo simple x, y and z movements (translations) will not be distorted by space and thus their dimensions will be maintained constant. We call such transformations 'invariant under translation'. This invariance property leads to the idea of conservation of linear momentum when we consider two inertial frames of reference. In a similar way, we think that space has no preferred directions, otherwise the results of experiments would depend upon the orientation of the apparatus. This property of isotropy leads to invariance under rotations and, in turn, is directly related to conservation of angular momentum.

Furthermore, we do not think that the results of experiment should depend upon the time at which they are carried out, and this concept is intimately related to conservation of energy. The basic symmetries of space and time which ensure that experimental results do not depend upon when and where they are carried out are obviously very fundamental. Indeed, it is debatable if we could have any physical science if results were unrepeatable and depended upon the whereabouts of the laboratory in which the measurements were made!

More generally, all symmetries, and hence invariances, are closely connected with corresponding conservation laws. In fundamental particle physics many new conservation laws have been discovered by observing new symmetries in the basic interactions. (Sometimes the reverse process has also occurred, where observed conservations have led to an understanding of an underlying symmetry.)

The transformations of space and time that we have just been discussing are 'continuous'. Rotations can be carried out through any arbitrarily small increment, as can translations in space and time. There are also 'discrete' symmetries in physical systems with corresponding discrete transformations.

A typical discrete transformation is that of inverting the spatial coordinates. This mirror transformation is brought about by the 'parity' operation which turns a vector $\mathbf{x}$ into a vector $-\mathbf{x}$. The question we are really asking ourselves is whether or not we could tell if we were looking at an experiment directly or via a mirror. In quantum mechanical terms (see Appendix B) the state of the system is defined by a wave-function $\Psi(\mathbf{x})$ and if we carry out the parity operation twice in succession we must come back to the state we started with.

Mathematically we can present these steps as follows:

$$P\ \Psi(\mathbf{x}) \rightarrow \Psi(-\mathbf{x}); \ P\ \Psi(-\mathbf{x}) \rightarrow \Psi(\mathbf{x})$$

where P is the parity operator, and so we must have $P^2 = 1$. If we now write the parity operation as

$$P\ \Psi(\mathbf{x}) = p\ \Psi(\mathbf{x})$$

where p is an ordinary number which represents the 'eigenvalues' of the operator P, then it is seen that $p = +$ or -1. We call a state with a well defined parity such that $p = +1$ an 'even' parity state and if $p = -1$ then it is an 'odd' parity state.

If an interaction is invariant under the parity operation, then the parity remains the same for all times, that is, it is conserved. In practice, it is found that for electromagnetic and strong interactions this conservation law seems to be exact. However, the weak force evidently does not conserve parity. Parity non-conservation was first observed in a historic experiment (in 1956) in which the beta-decay of cobalt-60 nuclei was studied by Chien Wu and Ernest Ambler. The cobalt was kept at a very low temperature and inserted in a magnetic solenoid in such a way that a large proportion of the nuclear spins were aligned in the magnetic field. The relative intensities of electrons emerging parallel and anti-parallel to the field were then measured. An asymmetry was observed corresponding to a certain degree of overall parity violation. The electrons were seen to be emitted preferentially on the side of the coil for which the current flows clockwise, that is, in the direction against the magnetic field. From the point of view of weak interactions, left and right can thus be distinguished in a mirror and the electrons must have predominantly a left-handed spin. These early results have since been confirmed in the case of the decay of the muon and in many other weak processes.

The very small parity-violating effects which are seen in apparently pure electromagnetic or strong interactions are actually manifestations of small admixtures of weak interaction effects. In some circumstances the weak force shows a preference for 'left-handed' processes and in others for 'right-handed' ones (see section 4.7).

A second discrete operation is that of time reversal, whose mathematical operator, T, has the effect of reversing all directions of motion, including spin. An invariance of an interaction under time reversal implies that, if some process occurs, then a second possible process must exist in which all directions of motion are reversed in the first process. Moreover, the rate of one process is calculable in terms of the other. Our everyday experience would suggest that this symmetry should be an absolute property of nature. Again, there is evidence in the weak interaction (in the decay of the long-lived neutral kaon, K) that this invariance principle is also violated to a very small degree.

We shall meet other conservation laws in later chapters ('charge conjugation' in section 3.2, 'strangeness' in section 4.2, 'lepton number' in section 4.7 and so on). The conservation laws that we have come across so far, together with an indication of their range of validity are summarized in Table 2.2.

Table 2.2. Conserved quantities.

Conserved quantity	Basic force		
	Weak	Electromagnetic	Strong
Energy	Yes	Yes	Yes
Linear momentum	Yes	Yes	Yes
Angular momentum	Yes	Yes	Yes
Electric charge	Yes	Yes	Yes
Baryon number	Yes	Yes	Yes
Isospin	No	No	Yes
Parity	No	Yes	Yes
Time reversal	Very small violation	Yes	Yes

CHAPTER 3

how do particles interact?

3.1. Fields and exchanged field particles

The word 'field' is used in physics to describe a physical parameter which is defined at all points in a certain region of space and time. In a static field the values of the parameters at every point do not change with time. If the physical quantity being described is a scalar, having a magnitude only, we refer to a scalar field. In classical physics, an example would be a scalar field describing the variation of temperature in a certain region of space.

A vector field is used to describe a quantity which has its magnitude and direction defined everywhere and therefore needs three components at every point in the region of space and time. A typical classical example appears in the consideration of fluid flow. The velocity field in a liquid defines the three components of the fluid flow at all points in the liquid.

In particle physics we are concerned with a number of fields. For example, particles carrying electric charge interact with other charged particles through the electromagnetic field. The modern theories of the fundamental interactions all involve the idea of an exchange of an object between the two interacting particles. It is assumed that all interactions between any two particles take place through the emission of a 'field particle' from the first particle and its subsequent absorption by the second particle. At the macroscopic level, a rugby player can interact with another player by throwing the ball to the second player who then catches it. The first player loses some momentum and the second player gains the same momentum increment. A similar type of process is envisaged at the microscopic level in interactions between the fundamental particles.

As an example, an electron with its electric charge can feel the presence of a second electron (that is, another repelling electric charge) by receiving photons emitted by the other electron. The photon is the quantum of light and the electromagnetic force theory is now understood in terms of the exchange of such quanta between electric charges.

Classically, if a force is applied to a charged particle it is accelerated and radiates an electromagnetic wave. In the field theory approach, the electromagnetic energy is emitted as a photon and the change of momentum that occurs in an interaction is produced by this emission. Such exchanges can be conveniently

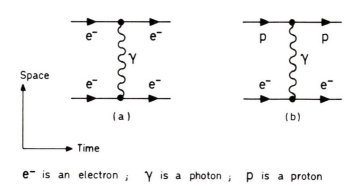

e⁻ is an electron ;   γ is a photon ;   p is a proton

Figure 3.1. Electromagnetic interactions (Feynman diagrams): (*a*) electron–electron and (*b*) electron-proton scattering.

represented using diagrams named after Richard Feynman (see Figure 3.1).

In Figure 3.1(*a*), the Feynman diagram shows the first electron scattering off the second one through the exchange of a photon. In Figure 3.1(*b*), the interaction is between a proton and an electron, again through a photon exchange process. In these diagrams time is depicted as flowing horizontally. The directions of the particles as they leave or enter the various vertices are shown by the directions of the arrow heads.

The nature of the exchange can perhaps be more easily understood if we first consider the nature of the electron itself. If we assume that emission and absorption of photons is responsible for the electromagnetic field, even in the context of a single electron, emission and subsequent reabsorption of a photon must be possible. Such a spontaneous emission of a photon would apparently violate the most important of all the physical laws, namely the conservation of energy. However, it is possible to violate this law by an amount of energy ΔE in a very short time interval Δt provided that $\Delta E.\Delta t \sim \hbar$. This is again an application of the uncertainty principle (see Appendix B.1). It arises from the idea that it would be impossible to detect, by any measurements whatsoever, a change of energy ΔE in a time Δt if the product is of the order of Planck's constant. The measurement of such a small energy cannot be made in such a short time, even in principle.

A single 'bare' electron is thus surrounded by a sea of so-called virtual photons (Figure 3.2). They are virtual in the sense that they are not free photons emitted as real observable quanta. It is perhaps now a little easier to understand

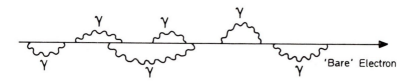

Figure 3.2. An electron 'dressed' with virtual photons.

the photon exchange in electron–electron scattering if we think of two 'dressed electrons' meeting each other. If the collision is close enough, the second electron can absorb the emitted photon before the first electron has been forced by the uncertainty principle to reabsorb it. It is as if two jugglers, throwing balls into the air, pass close by each other and one juggler catches one of the balls thrown into the air by the other. Thus the interactions shown in Figure 3.1 involve the exchange of virtual photons.

In some circumstances an electron can interact with another particle to form a bound state. For example, consider an electron colliding with a positron, a particle which has properties similar to those of the electron except that it carries a unit of positive charge. (The positron is the 'antiparticle' of the electron; see section 3.2.) In such a collision the electron may be simply scattered, but there is a finite chance that it will radiate a photon and form a bound state with the positron, producing an atom called positronium. This atom is chemically similar to hydrogen as it has just one valence electron, but it is about 2000 times lighter. It differs from hydrogen in another important respect in that it is not stable. Even if it is formed in a vacuum, a positronium atom has a definite lifetime before the electron and positron annihilate each other with the emission of electromagnetic energy in the form of two or three high-energy photons (γ-rays).

The picture of electromagnetic interactions just given leads to an understanding of both scattering processes and the existence of bound states. The quantitative understanding of the role of the exchange of virtual photons in electromagnetic interactions is the basis of the modern electromagnetic field theory. This theory incorporates the Maxwell equations, special relativity and quantum mechanics and is called quantum electrodynamics. It is the most complete and well tested theory in modern physics.

Many attempts have been made over the last two decades to use the same approach to set up field theories which would give an understanding of the other basic forces of nature. The exchanged field particles in the case of the weak force are the massive 'intermediate vector bosons' (Z^0, W^+ and W^-) which are analogous to the photons in the electromagnetic case. The large masses of these mediating particles cause the relatively feeble strength of the weak interaction and also severely limit the range of the weak force. The word 'vector' is used because the W and Z particles carry spin-1, which means that they have three spin substates (see Appendix B.4). They thus have three components and correspond to a vector field.

In important experiments carried out recently (1983), clear evidence for the production of these particles has been obtained by directly observing their decays (see chapter 5) and there is now great confidence that the weak interactions are being well described by the new theory which involves the exchange of the massive W and Z particles. Even more importantly, this new theory combines together the electromagnetic and weak forces into one scheme. This achievement is comparable in importance to the discovery by Maxwell that electric and magnetic forces could be combined within one framework.

In recent years, a candidate field theory for the strong interaction has also emerged, called 'quantum chromodynamics'. In this theory there are eight field particles called 'gluons' which are emitted and absorbed by the quarks. There is increasing evidence that this theory has most of the right elements in it and has good predictive power. Again, in recent experiments (1983) observations have been made which are interpreted as direct evidence for the existence of gluons (see chapter 6). Examples of weak, electromagnetic and strong scattering processes are shown in Figure 3.3.

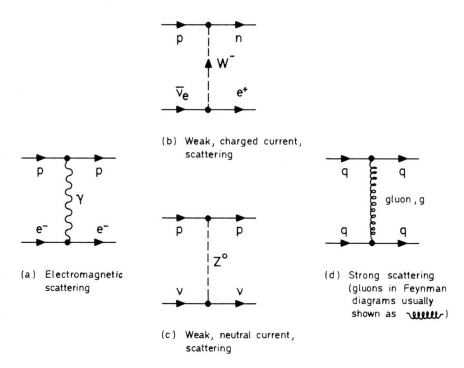

(b) Weak, charged current, scattering

(a) Electromagnetic scattering

(c) Weak, neutral current, scattering

(d) Strong scattering (gluons in Feynman diagrams usually shown as ⌇⌇⌇)

Figure 3.3. Examples of scattering processes.

For many years, it was thought that the pion was the basic particle that was exchanged in strong interactions as it had about the right mass to account for the range of the force (from an uncertainty principle argument). Although we now know that the pion is a composite particle, such meson exchange models can still be useful in providing a good description of some classes of strong interaction (see section 6.7).

Attempts are also being made to fit gravitation into the field theory framework. Although this problem will not be discussed in detail in this book, it is assumed that there is a field particle, the 'graviton', which is the gravitational analogue of the photon but has spin 2. The current list of exchanged field particles is summarized in Table 3.1.

Table 3.1. The exchanged field particles.

Force	Exchanged particles
Gravitation	Graviton
Weak force	Intermediate vector bosons: W^+, W^-, Z^0
Electromagnetic	Photon
Strong force	Gluons (set of 8)

Just as positronium and a hydrogen atom can be regarded as bound states (two charged particles held together by the electromagnetic interaction, that is, exchange of photons), the Earth and Moon constitute a bound state of the gravitational interaction and the pion (two quarks) or proton (three quarks) constitute bound states of the strong interaction. It is not possible to point to any system in which two particles are held together by the weak interaction (exchange of Z or W particles), but such an exchange is involved in the β decay of nuclei (see Figure 1.5).

3.2. Particles and antiparticles

In 1928, Dirac extended quantum mechanics and electromagnetism into the relativistic regime by considering a relativistic wave equation for an electron. He produced some surprising solutions which suggested the possibility not only of the expected two spin sub-states ('spin-up' and 'spin-down'), but also of 'negative energy' states for the electron. These states are now understood to correspond to the antiparticle of the electron, namely the positron already mentioned in section 3.1. The positron was predicted to have the same mass and spin angular momentum as the electron, but opposite electric charge (and magnetic moment). The evidence for this particle was first found in 1932 and the existence of positrons is now taken as commonplace. Certain artificially produced radioactive elements emit them.

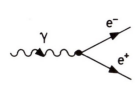

(a) Virtual process

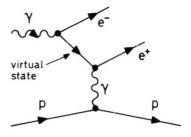

(b) Pair production in the Coulomb field of the proton

Figure 3.4. The pair production process.

A positron can also be 'pair-produced' in association with an electron in the interaction of a high-energy photon with matter, the energy of the photon being converted into mass in accordance with the relation $E = m_0 c^2$ (equation 2.3.3). This pair-production process cannot occur directly, as in Figure 3.4(*a*), since energy and momentum cannot be simultaneously conserved. It requires the coulomb field of the nucleus for the real production to take place, the nucleus then recoiling to take up the momentum transfer. Pair production in the electromagnetic field of the proton is shown in Figure 3.4(*b*).

The pair-production process must also be considered in our picture of the 'dressed' electron (see last section). The bare electron and some of its virtual interactions with the electromagnetic field can now be pictured as in Figure 3.5(*a*) and in a similar way the 'free' photon can be visualized as in Figure 3.5(*b*). (These diagrams involve the so-called self-energy terms (see Figure 3.5(c)), which technically introduce infinities into calculations of the bare mass and charge of the electron. However, a process known as renormalization has been devised which absorbs these infinities into the measured mass and charge of the real electron to give finite answers.)

Many other antiparticles have since been observed. For example, in 1955 the antiproton was discovered (by Chamberlain and colleagues) and more recently

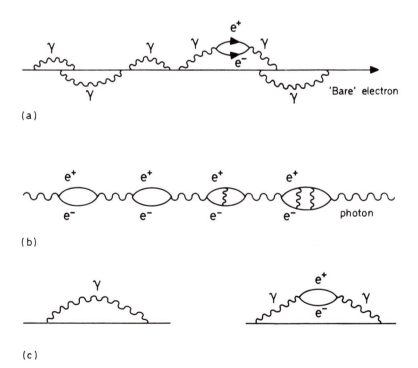

Figure 3.5. The 'free' (*a*) electron and (*b*) photon, and (*c*) the 'self-energy' terms.

even antideuterons have been observed. At the most fundamental level, we now believe that all particles have their antiparticle counterparts and their origin lies in the existence of families of antiquarks and antileptons (see chapter 4). An antiparticle is normally given the same symbol as its corresponding particle with a bar added over the top. Thus, the antiproton is written as $\bar{p}$.

The process of going from a particle to its own antiparticle corresponds to the theoretical operation, C, called 'charge conjugation'. This charge conjugation process has the effect of reversing the sign of the electric charge and the magnetic moment of a particle. For a baryon, charge conjugation also reverses the sign of the baryon number, B (and for a lepton it reverses the appropriate lepton number — see section 4.7).

In Feynman diagrams, particles going towards a vertex with a momentum $\mathbf{p}$ can be replaced by antiparticles coming from the vertex with momentum $-\mathbf{p}$, or vice versa. Provided the interaction considered is invariant under this charge conjugation process, then the C-operation may be applied to the whole diagram without changing the probability amplitude for the whole process depicted.

Experimentally, it is found that there is good evidence that the electromagnetic and strong interactions conserve charge conjugation. However, again the 'odd man out' is the weak interaction and C non-invariance is seen, for example, in the beta-decay process (see chapter 5).

Finally, while we are discussing antiparticles, mention should be made of one of the great cosmological mysteries in which particle physics has a crucial role. This is the question of why we see about us mainly particles, rather than a mixture of particles and antiparticles which would eventually result in just photons. Fermions and antifermions can only be created and destroyed in pairs, and it is not easy to understand how the symmetry of the situation has been lost. (Possibly, the fermion–antifermion pair rule might be broken at some level at the high energies which would have existed at the time of the 'Big Bang'.)

3.3. Cross-sections

In considering any interaction between two particles, we often speak of the cross-section for this particular process. Let us consider, for example, a beam of electrons fired at a target made of hydrogen. The nuclei of the hydrogen atoms are simply protons and most of the scattering observed will be due to the electromagnetic interactions between the individual incident electrons and these massive protons, as shown theoretically in Figure 3.3(*a*) and more practically in Figure 3.6.

Let there be n_e electrons incident on the target in unit time. Some electrons will pass through the hydrogen without interacting, while others will scatter off the protons through the coulomb force. Suppose that n_s electrons per second scatter, then the cross-section is defined as

$$\sigma = n_s/(n_e \times n_p) \qquad (3.3.1)$$

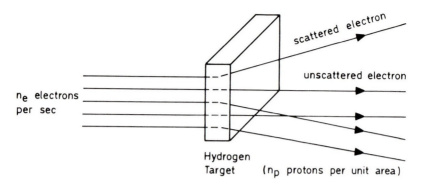

Figure 3.6. Electrons scattering off protons.

where n_p is the number of protons per unit area in the target. The dimensions of σ are those of area $(T^{-1}/T^{-1}L^{-2})$ and σ represents an effective area per target particle such that an incident particle hitting a disc of this area will be scattered. Thus the cross-section multiplied by the number of target particles per unit area gives the fraction of particles in the beam which will be scattered.

If we wish to write down the probability that a particle is scattered in a particular direction, then we use the 'differential cross-section', $d\sigma/d\Omega$, which is a measure of the scattering probability per unit solid angle about that direction.

Provided that we fully understand the nature of the interaction process, it must be possible to make a theoretical calculation of these cross-sections which can be checked by experiment. If we only have a model we can at least test it by experiments which essentially measure the probability for a particular scattering process to occur. Even when the theory is fully understood, as in the electromagnetic case, the full calculation is beyond the scope of this book. However, we can at least get a feel for the magnitudes involved by considering coupling constants, which are measures of the strengths of the various basic interactions.

In the case of the electromagnetic interaction, such as in the electron–proton scattering discussed above, the coupling constant α is called the fine-structure constant (as it enters into the calculation of the fine-structure line splitting in atomic spectra). This dimensionless constant α is given in SI quantities by

$$\alpha = e^2/4\pi\epsilon_0 hc \sim 1/137 \qquad (3.3.2)$$

where e is the charge on the electron (see Appendix A.2).

The electric charge couples to a photon with a probability amplitude of $\sqrt{\alpha}$. The square of this amplitude determines the probability for emission or absorption of a photon and this is proportional to e^2, from equation 3.3.2. Considering again the electromagnetic electron–proton scattering process shown in Figure 3.3(a) we see that the photon couples to both the charge on the electron and that on the proton. However, from relativistic kinematics (Appendix A.3) the photon must be considered to be virtual since it must possess a rest mass other than zero to convey a finite four-momentum transfer q between the electron and the proton. The scattering amplitude will involve a $\sqrt{\alpha}$ term for the electron–photon vertex, a further $\sqrt{\alpha}$ term for the photon–proton vertex and, in addition, a $1/q^2$ term called the propagation factor which is connected to the range of the coulomb potential. Hence, the cross-section σ, which is proportional to the overall scattering amplitude, is given by

$$\sigma \propto \left(\frac{\alpha}{q^2}\right)^2 \propto \frac{e^4}{q^4} \tag{3.3.3}$$

In carrying out a full electromagnetic calculation we must allow not only for single photon exchange, but also for two, three or more photons. The first- and second-order diagrams for e^-–e^- scattering are shown in Figure 3.7(a). The contributions of successive terms become less and less important as an extra power of α ($= 1/137$) comes into the amplitude for each additional exchange term. (On the other hand we must not include diagrams such as those shown in Figure 3.7(b), as they involve the 'self-energy' terms which have already been allowed for in the mass and charge of the electron through the renormalization process mentioned in the last section.)

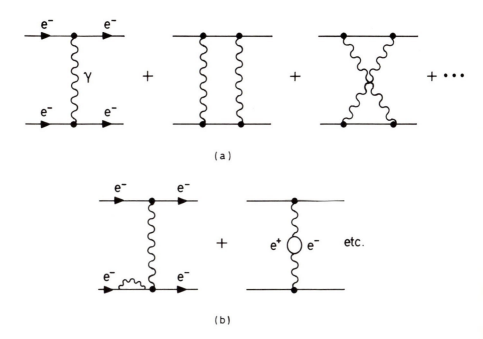

Figure 3.7. (a) First- and second-order e^-–e^- diagrams; (b) self-energy terms.

Typically one measures an electromagnetic cross-section of the order of a few tens of μb (see Appendix A.2 for note on units).

If we were to consider a weak interaction, such as that shown in Figure 3.3(*b*), then the cross-section takes the form

$$\sigma \propto \left(\frac{g^2}{g^2 + M_w{}^2}\right)^2 \qquad (3.3.4)$$

where g is an effective 'weak charge', by analogy with the electric charge, which couples to the intermediate vector boson, W. The extra term $M_w{}^2$ in the propagation factor arises from the non-zero mass of this exchanged field particle in this weak interaction case (see chapter 5). Weak interaction cross-sections are extremely small, typically 10^{-4} nb.

For a strong interaction, such as that between two quarks depicted in Figure 3.3(*d*), the situation is rather more complicated, as we shall see in chapter 6. The characteristic coupling, α_s, is not in fact constant but depends on the momentum transferred in a rather dramatic way. For small momentum transfers which probe the proton over distances of the order of the size of the whole proton (~ 1 fm), α_s has a value of the order 1. Thus the coupling in this condition is about 100 times stronger than in the electromagnetic case. On the other hand, as the momentum transfer increases, corresponding to the distance between the quarks getting smaller, the value of α_s decreases until the quarks behave as if they are 'free' at extremely short distances. We can picture such a coupling as if the quarks were connected together with elastic such that the force between them would increase as we try to stretch them apart but leaves the particles nearly free when the rubber is not extended. The parameter α_s is often called a running coupling constant. The total cross-sections involved in strong interaction processes have a magnitude of the order of tens of millibarns.

We shall go into these questions in somewhat more detail in the later chapters of this book.

The idea of constituents of baryons was finally confirmed when direct evidence was obtained in a series of experiments which probed the nucleons with high-energy electrons and neutrinos. The huge linear accelerator at the Stanford Linear Accelerator Center, California, was used to probe protons with 20 GeV electrons. This experiment explored how the electric charge (together with the magnetic effects due to any spinning charged constituents) was distributed within the proton. Had the charge been spread uniformly throughout the whole volume of the proton, large numbers of very small-angle scatters would be seen. If, on the other hand, the charge is centred on very small constituents such as quarks, then when the probing electron happens to come very close to a quark (that is, a small 'impact parameter') there will often be a dramatically large-angle scatter. The analysis of this scattering experiment showed that there were indeed a large number of wide-angle scatters which were seen to be characteristic of point-like scattering centres within the target nucleon. (Exactly the same effect, but on a larger scale, was seen by Rutherford when his scattering experiments established the relatively small size of the nucleus within the atom as a whole.)

Similar scattering experiments on the proton were carried out in CERN, Geneva, but neutrinos were used as the probing particles. These neutrinos feel only the weak force (see section 1.5 and chapter 5) and it is therefore the distribution of 'weak charge' within the proton which is being explored. Again, discrete scattering centres were confirmed. These experiments also gave a hint that there were, in addition to the charged quarks, further electrically neutral constituents (later identified as the gluons — see chapter 6.) Finally, experiments carried out at the CERN intersecting storage rings proton–proton collider (see Figure 2.1) again produced large numbers of wide-angle scatters which corresponded to individual quark–quark head-on collisions.

As quarks carry fractional charges, if they exist as separate entities they would interact with matter in a very characteristic way. (The process of energy loss through ionization, that is, knocking electrons out of atoms leaving positively charged ions, depends on the square of the electric charge carried by the fast particle.) Also, it is possible in principle to measure the charge of an isolated quark in an experiment similar to that of Millikan which established the unit charge carried by electrons. However, no firm evidence has been obtained in either type of experiment that establishes without doubt the existence of free quarks. It seems that they probably exist only in bound states confined within the hadrons. (This is a topic that we shall return to in the discussion of the strong interaction in chapter 6.)

CHAPTER 4

the basic 'building blocks' of matter

4.1. The quarks

The strongly interacting particles, such as protons, neutrons, pions and kaons, are now understood in the current so-called standard model as consisting of systems of quarks. These individual and presumably indivisible building blocks of matter exist in at least five, and probably six, basic kinds usually called flavours. These quarks all carry fractional electric charge, either $+2/3$ or $-1/3$ of the magnitude of the charge on the electron. All the quarks carry spin 1/2 and are therefore fermions (see section 2.4). Five of these flavours of quark (up, down, strange, charm and beauty or bottom) have been established but as yet only preliminary evidence exists for the expected sixth flavour (top). In this scheme, each type of quark has a corresponding antiquark (see section 3.2). The antiquarks are usually depicted by a barred letter, e.g. $\bar{u}$ for the anti-up quark. The complete set of quarks is shown in Table 4.1.

Table 4.1 The quarks and antiquarks.

Quarks			Antiquarks		
Charge	Spin	Flavour	Charge	Spin	Flavour
$+2/3$	1/2	Up (u), Charm (c), Top (t)	$-2/3$	1/2	$\bar{u}$, $\bar{c}$, $\bar{t}$
$-1/3$	1/2	Down (d), Strange (s), Beauty (b)	$+1/3$	1/2	$\bar{d}$, $\bar{s}$, $\bar{b}$

With these six flavours of quarks, and the corresponding antiquarks, one can make up all the known hadrons that previously had been thought of as elementary, such as the proton and pion. There are actually hundreds of such hadrons which are all quark composites and are clearly not elementary in the usual sense of the word. Many of these hadrons have been discovered among the debris of the products of the collisions of two initial hadrons. The typical size of these composites is of order 1 fm (10^{-13} cm) while the individual quarks are $< 10^{-3}$ fm and thus appear to be essentially point-like within the relatively large composite particles.

The evidence for such constituents has come partly from the symmetries observed in the 'elementary' particle properties. Until 1974, the only observed hadrons were composed of u, d and s quarks. The symmetry that was observed came from seeing the patterns of particles produced when families of particles with the same spin were grouped together. The patterns produced related to two variables. The first property is the electric charge carried by each particle and the second property we now recognize as relating to the s quark content of the particles. These patterns noticed by Yuval Ne'eman and Murray Gell-Mann were called the 'eightfold way'. They reflect the underlying symmetry, related to the mathematical group SU(3), which eventually led to the hypothesis of quark constituents. This suggestion of quarks came independently from George Zweig and Murray Gell-Mann, but at this stage they were limiting themselves to three quarks only (u, d and s). At the time, the idea was not received well as there was great resistance to the concept of fractionally charged constituents.

4.2. The quark structure of the simplest hadrons

Particles which feel the strong interaction, the hadrons, form two main groups. The baryons are formed from any combination of three quarks and the mesons consist of a quark and antiquark pair (not necessarily of the same kind). As mentioned before in section 2.5, the baryons are fermions but the mesons are bosons.

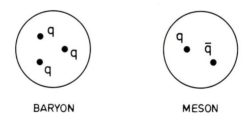

BARYON   MESON

(q stands for any quark flavour)

Figure 4.1. Quark structure of hadrons.

For example, a proton consists of a combination of u, u and d quarks giving a total electric charge of $(2/3 + 2/3 - 1/3) = +1$ units of charge. On the other hand the neutron consists of d, d and u quarks with a total charge of $(-1/3 - 1/3 + 2/3) = 0$ units of charge. The positively charged pion in this picture consists of a $(u\bar{d})$ combination with an electric charge of $(2/3 + 1/3) = +1$ unit, whereas the negative pion is obtained from a $(d\bar{u})$ pair with a combined charge of -1 units.

When three quarks each with spin 1/2 are combined together, bearing in mind the vector nature of angular momentum, the spin angular momentum states

1/2 and 3/2 are both possible. In the simplest case where we assume there is no relative angular momentum involved, the combined systems can have a total angular momentum of 1/2 or 3/2. Thus the (u u d) state can have spin 1/2, and represents the proton, or can have spin 3/2. In this latter case, the particle has been identified as the Δ^+. Similarly the Δ^0 is the (d d u) spin 3/2 analogue of the (d d u) spin 1/2 neutron. We would therefore expect one 'ground state' family of baryons for each of the possible lowest spin states.

A similar picture emerges for the mesons, where the quark and antiquark each with spin 1/2 can combine in spin 0 or spin 1 states. Thus the (u d̄) and (d ū) pairs in spin state 0 correspond to the positive and negative pions, respectively, but in the spin 1 state give rise to two more particles, the positive and negative rho-mesons. Table 4.2 summarizes the quark structures of the particles so far discussed.

Table 4.2. Some simple hadron structures.

Quark combination	Charge	Spin	Particle	
u u d	+1	1/2	Proton	
		3/2	Δ^+	
d d u	0	1/2	Neutron	
		3/2	Δ^0	
u d̄	+1	0	+ve pion	π^+
		1	+ve rho	ϱ^+
d ū	−1	0	−ve pion	π^-
		1	−ve rho	ϱ^-

If we now consider the strange quark, s, we will get further possible baryon states with one, two or three s quarks. The inclusion of an s or s̄ antiquark in a meson will give rise to many further meson states. (Notice that the most familiar and abundant particles, such as the proton, neutron and pion, require only the up and down flavours.) The particle states containing an s quark are found to be typically $150 \, \text{MeV}/c^2$ heavier than their non-strange counterparts.

Considering the first three flavours only (u, d and s), and taking zero relative orbital angular momentum states only, the ground state spin 1/2 and 3/2 baryon families have eight and ten members, respectively, as shown in Figure 4.2.

One of the triumphs of Murray Gell-Mann's ideas of the symmetry of the hadrons was the prediction of the existence and properties of the baryon consisting of three s quarks. In 1963 this previously unobserved particle was found with just the right properties.

In early discussions of the properties of the composite particles, reference is often made to the 'strangeness' of a particle. We now understand 'strangeness', S, as simply meaning the negative of the number of s quarks in the state. Thus, each s quark carries $S = -1$ and we assign $S = +1$ to each s̄ quark. This definition, with its rather arbitrary sign convention, arose historically from the observation

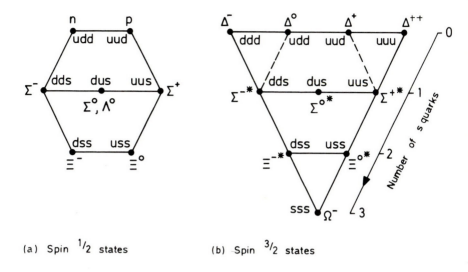

(a) Spin $^1/_2$ states   (b) Spin $^3/_2$ states

Figure 4.2. First octet and decuplets of baryons.

that a particular class of particles, now known to be those that contain s and $\bar{s}$ quarks, could only be produced in pairs in strong interactions. An example of this process, known as associated production is

$$\pi^+ + n \rightarrow \Lambda^0 + K^+$$

In the quark model we now see this process as

$$(u\bar{d}) + (ddu) \rightarrow (dus) + (u\bar{s})$$

We see that the flavours are balanced on both sides of the equation and in particular we see that the Λ^0 has $S = -1$ and the K^+ has $S = +1$. This is an example of conservation of strangeness in the strong interaction. (The threefold symmetry displayed in Figure 4.2 is connected with such conservation laws.)

In Figure 4.2 we see that families of particles with either one, two, three or even four charged states occur. We must therefore extend our ideas of isospin, introduced in section 2.5, to allow for all these possibilities. Figure 4.3 shows the values of I, I_z and S for the complete spin 3/2 decuplet.

Such patterns as those shown in Figures 4.2 and 4.3 demonstrate the 'eightfold way' symmetry mentioned in the last section. We can now understand the origin of these symmetries in terms of the quark content of the particles. The properties of the basic mathematical group reflecting this symmetry, SU(3), correspond to the rules for building up the composite particles from the three types of constituents, u, d and s quarks.

Now we have introduced the third quark flavour (S), we must also modify

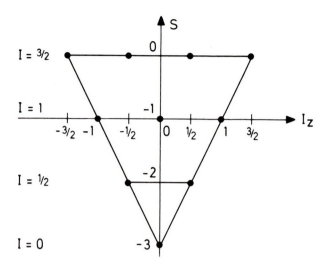

Figure 4.3. I, I_z and S for spin 3/2 baryon decuplet.

equation 2.5.2 to compute the charge carried by a particle as follows:

$$Q = I_z + (B + S)/2 \qquad (4.2.1)$$

where a baryon number B of unity is now given also to the hyperons. The previous relation (2.5.2) holds only for non-strange quark combinations. As an example, consider the Σ^+ which has $I_z = 1$, $S = -1$ and $B = 1$ and therefore has

Table 4.3. The first nonet meson family.

Quark combination	Charge	Spin	Particle
u d̄	+1	0	π^+
		1	ϱ^+
u ū d d̄ s s̄	0	0	π^0, η^0, η^1
		1	$\varrho^0, \omega^0, \phi^0$
d ū	−1	0	π^-
		1	ϱ^-
u s̄	−1	0	K^+
		1	K^{*+}
d s̄	0	0	K^0
		1	K^{*0}
s ū	−1	0	K^-
		1	K^{*-}
s d̄	0	0	$\overline{K^0}$
		1	$\overline{K^{*0}}$

$Q = 1 + (1 - 1)/2 = +1$ whereas the X^- has $I_z = -1/2$, $S = -2$ and $B = 1$ and so has $Q = -1/2 + (1 - 2)/2 = -1$ (see Figure 4.2(a) for the quark content of these states).

Still using only the first three flavours, we can construct $(3 \times 3) = 9$ q$\bar{q}$ pairs, each combination having spin 0 or 1 in the lowest energy states. Table 4.3 shows the complete set of states in which there is zero relative orbital angular momentum between the quark and the antiquark. These states with $L = 0$ are referred to as 'S-wave states', following the spectroscopic nomenclature.

4.3. Excited quark states

Using the first three flavours of quark only (u, d and s), it is possible to generate a vast number of additional 'fundamental'(?) particles by considering the situation where there is a non-zero relative orbital angular momentum between the constituent quarks. This will correspond to having excited states of quarks with higher angular momentum and energy (mass) which will then decay to lower mass states, very often with the emission of pions. This is exactly analogous to excited atomic states with high angular momentum and energy which then drop back to the ground state with the emission of photons of definite energies. Thus, many families of generally more massive baryons and mesons are predicted with relative orbital angular momenta, $L = 1, 2, 3 \ldots$ corresponding to P, D, F . . . waves in the spectroscopic notation.

As an example, consider a q$\bar{q}$ pair with $L = 1$, so that the total angular momentum $\mathbf{J} = \mathbf{L} + \mathbf{S}$. Hence we have $L = 1$ orbital angular momentum and spin angular momentum $S = 0$ or 1. With $L = 1$ and $S = 0$ we only have $J = 1$ states, which are referred to as $^1P_1$ states (where the notation is $^{2s+1}P_J$). In the second case with $L = 1$ and $S = 1$ there are three possible values for J, that is, 0, 1 and 2. These states are called $^3P_0$, $^3P_1$ and $^3P_2$, respectively. Many examples of these excited states have been identified experimentally.

Such excited meson states, and their baryonic counterparts, usually have a considerably higher mass than their ground-state analogues and often decay very rapidly (in 10^{-23} s) to the ground states with the emission of pions. For example, some of the excited states of the negatively charged pion (d$\bar{u}$) are listed in Table 4.4 and it will be noticed that the mass increases with the higher excitations. What we see in these excited states are complete replicas of the ground-state families (or 'multiplets') of baryons and mesons of a given spin. Each ground state multiplet has a corresponding set of excited states just as we saw for the case of the negative pion in Table 4.4.

There is a remarkable agreement between the observed particle states and the predicted baryons and mesons expected with the ground and excited states in this simple quark model. So far, no obviously 'exotic' states have been observed that do not fit in with the simple 3q or q$\bar{q}$ model, although there are many predicted states that have not yet been observed. These omissions are usually attributed to the difficulty of separating and observing these states rather than to any belief that the states do not exist.

Table 4.4. The negative pion and some of its observed excited states.

Particle	S	L	J	Mass (MeV/c^2)	Strong decay	Width (MeV)
π^-	0	0	0	140	—	
ϱ^-	1	0	1	770	$\pi^- \quad \pi^0$	154
δ^-	1	1	0	983	$\eta \quad \pi$ $\quad\hookrightarrow \pi \ \pi \ \pi$	54
A_1^-	1	1	1	1275	$\varrho \quad \pi$ $\quad\hookrightarrow \pi \ \pi$	315
A_2^-	1	1	2	1318	$\varrho \quad \pi$ $\quad\hookrightarrow \pi \ \pi$	110
A_3^-	0	2	2	1680	$f \pi$ or $\varrho \pi$ $\quad\hookrightarrow \pi \ \pi \rightarrow \pi \ \pi$	250

4.4. The 'heavy flavour' quarks

Until 1974, all the observed families of 'elementary' particles could be understood in terms of combinations of just three flavours of quarks (u, d and s) and their antiquark partners. However, in that year a new very heavy meson (with a mass of $3100 \, \text{MeV}/c^2$) was discovered in two experiments, one led by Burton Richter at Stanford in California using an electron-positron collider ('SPEAR'), and the other led by Samuel Ting at the Brookhaven Laboratory with a conventional proton accelerator.

The evidence for this new particle implied the existence of a fourth flavour, charm (c). The new meson (called the ψ) was observed to have a very narrow width (about 63 keV) and is therefore rather stable. Typical strong decays occur in times of the order of 10^{-24} s with corresponding widths of the order of hundreds of MeV. How can one explain the relative stability of the ψ? The answer seems to lie in the fact that it consists of a $q\bar{q}$ pair, where the quark carries a new flavour ψ (charm) which is conserved in strong interactions. The antiquark is a $\bar{c}$ and it is the lowest mass state of the $c\bar{c}$ type.

Let us first consider the ϕ-meson which consists of an $s\bar{s}$ pair. This meson has a mass of $1019 \, \text{MeV}/c^2$ and can decay relatively fast (with about a 4 MeV width) into a positive kaon and a negative kaon each of which has a mass of $494 \, \text{MeV}/c^2$ through the mechanism shown in Figure 4.4(a). Here the s quark goes into the K^- meson and the $\bar{s}$ antiquark goes into the K^+ meson. Infrequently, it can also decay through an 'internal' annihilation of the $s\bar{s}$ pair into three gluons which then materialize into a set of non-strange hadrons, see Figure 4.4(c). (This annihilation decay is exactly analogous to the electromagnetic decay of one of the states of positronium, see section 4.5 below.) Such an annihilation process is suppressed

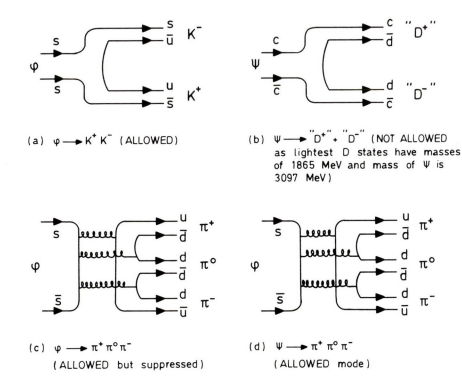

(a) $\varphi \longrightarrow K^+ K^-$ (ALLOWED)

(b) $\psi \longrightarrow \text{"}D^+\text{"} + \text{"}D^-\text{"}$ (NOT ALLOWED as lightest D states have masses of 1865 MeV and mass of ψ is 3097 MeV)

(c) $\varphi \longrightarrow \pi^+ \pi^0 \pi^-$ (ALLOWED but suppressed)

(d) $\psi \longrightarrow \pi^+ \pi^0 \pi^-$ (ALLOWED mode)

Figure 4.4. Decays of the ϕ-meson and allowed and forbidden decays of the ψ-meson.

compared to the case where the strange quark and antiquark survive into the final state.

In the case of the $\psi(c\bar{c})$, there are evidently no states with a single c or $\bar{c}$ quark with a low enough mass to allow a strong decay of the type shown in Figure 4.4(b). Instead the decay must go through the relatively suppressed mechanism which involves the internal annihilation of the c and $\bar{c}$ quarks and coupling to the decay particles through gluons as in Figure 4.4(d). For this reason the ψ is comparatively stable and so lives for a relatively long time.

In fact, a considerable number of meson and baryon states containing single c and $\bar{c}$ quarks have been discovered, but the lowest mass particle is the D-meson with a mass of $1865\,\text{MeV}/c^2$. Table 4.5 shows some of the observed charmed states. As an example, the charmed analogue of the strange K^--meson ($s\bar{u}$) is called the $D^0(c\bar{u})$. It is clear that introducing a fourth quark flavour, c, is going to increase enormously the number of expected states. For example, let us again consider the ground-state multiplet of spin 3/2 baryons as shown in Figure 4.2(b). With the fourth flavour we will now expect not the simple triangular pattern shown, but a tetrahedral diagram (see Figure 4.5) in which the number of c quarks in each state is shown increasing in the vertical direction.

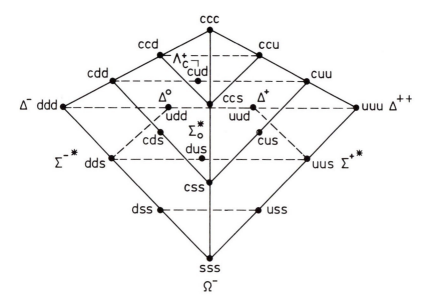

Figure 4.5. Spin 3/2 baryon decuplet with u, d, s and c quarks ('supermultiplet').

Table 4.5. Some observed 'charmed' mesons.

Quark combination	State	Mass (MeV/c^2)
$c\bar{u}$	D^0	1865
$c\bar{d}$	D^+	1869
$c\bar{s}$	F^+	1975
$\bar{c}u$	$\bar{D}^0$	1865
$\bar{c}d$	D^-	1869
$\bar{c}s$	F^-	1975
$c\bar{c}$	ψ	3097

The lowest mass charmed states are relatively stable and have a very long lifetime (by nuclear standards!) of the order of 10^{-12}–10^{-13} s. It is clear that their eventual decays are going through the weak interaction with a characteristically long timescale. It will be seen later that only through this process can the conservation laws for quark flavours be broken.

Charmed particles that have sufficient energy to be travelling at velocities comparable with the speed of light can survive for measurably long distances. For example, a particle with a mean lifetime of 10^{-12} s travelling with a γ of 10 (see Appendix A.3) will go a mean distance of about 3 mm before decaying. By using suitable high-resolution particle detectors, such distances can be observable directly, so that the details of the decay vertex can be studied. One useful high-resolution detector frequently used in these studies consists of blocks of photographic emulsion in which the inherent spatial resolution is about 1 micron (10^{-6} m). Such 'nuclear emulsions' are often used as targets for studying the production and subsequent decay of relatively long-lived states such as charmed particles. Frequently, an external detector is used to point back into the emulsion to the region of the production and decay vertices whenever these external detectors record secondary particles with a signature that suggests the possibility of charmed particle decay.

As an example, Figure 4.6 shows such an event in an emulsion experiment where incident neutrinos were used to produce the charmed particles. The high-energy neutrinos were fired at blocks of emulsion placed close to the window of a large bubble chamber. The observed secondary tracks in the bubble chamber were used to trace back to the interaction vertex in the emulsion where the subsequent decay could be studied in great detail. Figure 4.6 shows the first example ever observed of the charmed baryon Λ_c^+ which consists of a (cud) combination of quarks (see Figure 4.5). In this particular event the charmed baryon is seen to decay in the mode

$$\Lambda_c^+ \rightarrow p + K^- + \pi^+$$
$$(\text{cud}) \rightarrow (\text{uud}) + (s\,\bar{u}) + (u\,\bar{d}) \qquad (4.4.1)$$

where all the final state particles were uniquely identified in the bubble chamber. The flight time of the particle in this event was 7×10^{-13} s. (Note that in this decay the charm quark has disappeared and a strange quark has appeared. We will see later that this is typical in charm decays.)

A further extension to the simple quark model occurred in 1978 when an even heavier new meson called the upsilon, Υ, was discovered with a mass of $9 \cdot 5\,\text{GeV}/c^2$. This was found in an experiment at FNAL by a team led by Leon Lederman in which protons were fired at heavy nuclei and resulting pairs of oppositely charged muons were observed. An analysis of the observed muon energies showed clear evidence for the decay of a new particle with a very high mass (around $9 \cdot 5\,\text{GeV}/c^2$).

Table 4.6. Beauty meson states.

Quark combination	Charge	Particle
$b\,\bar{u}$	-1	B^-
$b\,\bar{d}$	0	B^0
$b\,\bar{s}$	0	B_s^0
$b\,\bar{c}$	-1	
$\bar{b}\,u$	$+1$	B^+
$\bar{b}\,d$	0	$\bar{B}^0$
$\bar{b}\,s$	0	$\bar{B}_s^0$
$\bar{b}\,c$	$+1$	
$b\,\bar{b}$	0	Υ

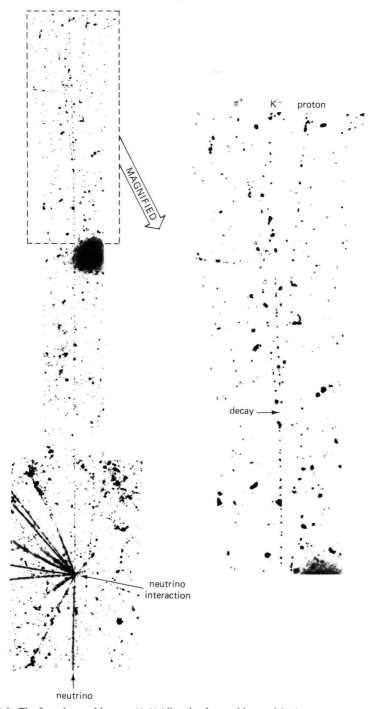

Figure 4.6. The first charmed baryon (Λ_c^+) (directly observed in emulsion).

Yet another 'heavy flavour', B, must be postulated to understand this new particle. This new flavour was given the name 'beauty' (or sometimes, rather less elegantly, it is known as 'bottom'!) This b quark and the corresponding anti-quark, $\bar{b}$, lead us to expect another set of massive particles such as the mesons shown in Table 4.6.

4.5. Quarkonia

In chapter 3, the 'positronium' atom was briefly discussed. This atom, consisting of a bound state of one electron and its antiparticle the positron, is an electromagnetic analogue of a strongly bound quark–antiquark pair. The positronium system has been extensively studied both theoretically and experimentally and it is therefore worth a little digression to discuss this interesting atom as a model for our qq states.

Positronium can be formed with the electron, positron spins parallel ('ortho-positronium') or antiparallel ('para-positronium'). The $e^+ e^-$ pair eventually annihilate into three or two photons for the ortho- and para-states, respectively. The lifetime of the ortho-state in vacuum is about 10^{-7}s, whereas the para-state has a natural lifetime of only 10^{-10}s. Positronium thus consists of a fermion-antifermion pair with a relative orbital angular momentum L and a spin angular momentum S ($=0$ or 1). Thus the lowest few states can be labelled as in Table 4.7 and the corresponding observed energy-level diagram is given in Figure 4.7(a).

Table 4.7. Quantum numbers of $e^- e^+$ or $q \bar{q}$ pairs.

L	S	Spectroscopic notation
0	0	$^1S_0$
	1	$^3S_1$
1	0	$^1P_1$
	1	$^3P_0$
		$^3P_1$
		$^3P_2$
2	0	$^1D_2$
	etc.	

A very close analogy obviously exists between the positronium bound state $e^+ e^-$ and the various bound states of $q \bar{q}$. In particular, a system consisting of the $c \bar{c}$ state, such as the ψ, has very similar properties to positronium and is often referred to as 'charmonium'. For example, just as ortho-positronium annihilates into three photons, so the corresponding charmonium state (which is the ψ) decays through the $c \bar{c}$ annihilation into three gluons, as in figure 4.4(d).

Many of the lower states of the $c \bar{c}$ system have now been identified and their mass (energy) levels are indicated in Figure 4.7(b). The similarity with the positronium levels in Figure 4.7(a) is striking (albeit on a very different energy scale).

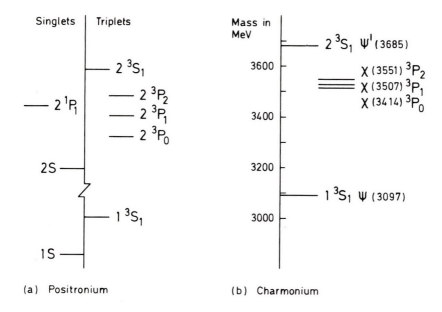

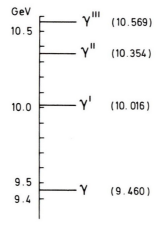

Figure 4.7. Energy levels for (*a*) positronium and (*b*) charmonium.

Figure 4.8. Observed $b\bar{b}$ states.

A good fit to the charmonium energy levels can be obtained by assuming that there is a binding potential between the quarks which increases linearly with distance.

Again, in direct analogy, the $b\bar{b}$ system has also been studied and a number of states observed (see Figure 4.8) which are referred to as 'beautyonium'. In general, such $q\bar{q}$ systems are called 'quarkonia'.

4.6. Coloured quarks

The quarks, being spin 1/2 particles, are fermions and must obey the Pauli Exclusion Principle. (see section 2.4.) There is an obvious problem with regard to states of three identical quarks, such as the particle Ω^- consisting of three s-quarks with zero relative orbital angular momentum. How can such a state exist without violating the exclusion principle? The answer lies in the idea that the quarks have another property that has so far not been discussed.

Oscar Greenberg suggested that quarks carry a new kind of 'charge' called colour. Each flavour of quark can exist in one of three colour states, referred to as blue, green and red. There is thus an additional quantum number which can be different for each of the s quarks in the Ω^-. Hence, we can avoid violating the exclusion principle. This idea is much more than a mere trick to get over the difficulty of the state of three identical quarks. It is, in fact, the basis of the modern understanding of the strong interaction and will be discussed in more detail in chapter 6.

It also appears to be the case that all observable particle states are 'colour singlets' that is, they are uncoloured, or 'white'). Thus, a baryon must contain one red, one green and one blue quark in the appropriate flavours to make up the other required properties. On the other hand, a meson will consist of a quark of one particular colour (e.g. red) and an antiquark carrying the anticolour charge (in this case, anti-red).

In a strong interaction, the exchanged gluons must themselves be coloured and in the course of such an exchange the quark colours will be changed (see Figure 4.9). However, the final state observed particles are always colourless. The field theory which attempts to describe the coloured gluon exchange mechanism as the basis of the strong force is called quantum chromodynamics (see chapter 6). One can already see from what has been said so far that an essential difference between quantum electrodynamics and quantum chromodynamics is that whereas the photon carries no electric charge, the gluons do carry the colour

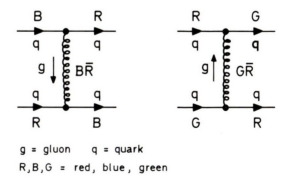

g = gluon   q = quark
R,B,G = red, blue, green

Figure 4.9. Examples of gluon exchanges between coloured quarks.

charge. It is thought (but not yet proved) that this property of the gluons can account for the fact that quarks are 'confined' within the particle dimensions, which is another way of saying that only colourless objects are observable.

4.7. The leptons

The particles which feel the weak force but not the strong force are referred to as leptons. The complete family of all known leptons is shown in Table 4.8. The charged leptons have electromagnetic as well as weak interactions but the neutral leptons (the neutrinos) are unique in that they feel only the weak force. These neutrinos are therefore very penetrating and, with modest energies of a few GeV, have an exceedingly long mean free path between collisions (typically of the order of 10^{20} cm of earth!). In addition, there are the corresponding antiparticles in the lepton family consisting of the three varieties of antineutrino and the positively charged counterparts of the electron, muon and tau particles.

Table 4.8. The lepton family.

Lepton		Charge	Mass
ν_e,	electron neutrino	0	
ν_μ,	muon neutrino	0	
ν_τ,	tau neutrino	0	
e^-,	electron	-1	$0 \cdot 511$ MeV$/c^2$
μ^-,	muon	-1	$105 \cdot 6$ MeV$/c^2$
τ^-	tau	-1	$1 \cdot 87$ GeV$/c^2$

The neutrinos are neutral particles of negligibly small rest mass and are fermions with half-integral spin. Interactions involving the leptons occur in such a way as to conserve separately three quantum numbers, one for each of the massive lepton types. For example, the e^- has electron lepton number $+1$, the antiparticle (positron) is assigned to the number -1, the ν_e has electron lepton number $+1$ and the $\bar{\nu}_e$ has the number -1. Thus, if we consider the decay of the $\overline{\text{muon}}$, μ^+:

$$\mu^+ \rightarrow e^+ + \nu_e + \bar{\nu}_\mu \quad (4.7.1)$$

Electron lepton number: $\quad 0 \rightarrow (-1 \quad +1 \quad +0)$
Muon lepton number: $\quad -1 \rightarrow (\ 0 \quad +0 \quad -1)$

In this example we see both the electron and muon lepton numbers being conserved. Another example is beta-decay of an atomic nucleus in which a neutron transforms to a proton with the emission of an electron and a neutrino as follows:

$$n \rightarrow p + e^- + \bar{\nu}_e \quad (4.7.2)$$

Electron lepton number: $\quad 0 \rightarrow (\ 0 \quad +1 \quad -1)$

Again we see the lepton number conservation. On the other hand, we do not observe decays such as

$$\mu^+ \rightarrow e^+ + \gamma \qquad (4.7.3)$$

which would break the lepton number rule.

Neutrinos have a curious property in that they are always found to be longitudinally spin-polarized. That is to say that if their velocity vector lies along the z-axis, the z-component of spin angular momentum will be $S_z = -1/2$. Thus the neutrino moves along as a 'left-handed screw' which is more formally referred to by saying that it has a 'helicity' of -1. On the other hand, the antineutrino is right-handed and has a helicity of $+1$. (These facts relate to the statement that the weak interaction violates parity, P, and charge conjugation, C, invariance.)

4.8. Summary — the basic constituents in the 'standard model'

We can now summarize the list of fundamental building blocks that we have encountered and which currently constitute the most elementary particles which are recognized in the so-called 'standard model'. There are two main groups of fermions: the quarks and the leptons. Each of these groups seems to exist in three 'generations' of pairs of particles which have masses that increase with generation number. We also have the field bosons discussed in the first chapter. Table 4.9 gives the complete list of the fundamental constituents, together with the antiquarks and antileptons.

Table 4.9. The lepton family.

	Generation	-1	$-2/3$	$-1/3$	0	$+1/3$	$+2/3$	$+1$
					Charge			
Quarks	1			d			u	
	2			s			c	
	3			b			t(?)	
Leptons	1	e			ν_e			
	2	μ^-			ν_μ			
	3	τ^-			ν_ϱ			
Antiquarks	1		$\bar{u}$			$\bar{d}$		
	2		$\bar{c}$			$\bar{s}$		
	3		$\bar{t}(?)$			$\bar{b}$		
Antileptons	1				$\bar{\nu}_e$			e^+
	2				$\bar{\nu}_\mu$			μ^+
	3				ν_τ			τ^+

Field bosons
3 gravitons, W^+, W^-, Z^0
1 photon, γ
8 gluons

CHAPTER 5

the electroweak interaction

5.1. The weak force — charged currents

Weak forces act between leptons and quarks, leptons and leptons, and quarks and quarks. When quarks are involved in the process, flavour changing can occur (see section 5.4), unlike the strong interaction in which flavours are conserved. It seems that only 'left-handed' particles and 'right-handed' antiparticles carry the weak charge, the opposite members such as right-handed electrons playing no part in the weak interaction. Quark–quark interactions are usually dominated by their strong interactions but the flavour-changing processes betray the presence of the weak interaction.

Until 1973 the body of data from weak interaction phenomena included beta-decays, weak decays of such particles as the pion, muon and particles containing s quarks, and neutrino interactions with nucleons. All of the observed processes were known to be 'charged current' interactions. This means that the force is mediated by electrically charged particles which we call the W^+ and the W^-. All the processes were characterized by a very small coupling constant which Enrico Fermi had written as

$$G_F \simeq 1\cdot02 \times 10^{-5}/m_p{}^2 \tag{5.1.1}$$

(in natural units)

Beta-decay of nuclei with either electron or positron radioactive emission involves weak transformations between protons and neutrons in the 'semi-leptonic' processes.

$$p \rightarrow n + e^+ + \nu_e \tag{5.1.2}$$

$$\text{and } n \rightarrow p + e^- + \bar{\nu}_e \tag{5.1.3}$$

Such decays can occur only in nuclei where the difference in mass energies of the parent and daughter nuclei make them kinematically possible. Note that the baryon and electronic lepton numbers are conserved. These reactions are related to the processes which correspond to the weak interactions of leptons with nucleons:

$$\bar{\nu}_e + p \rightarrow n + e^+ \tag{5.1.4}$$

$$\text{and } \nu_e + n \rightarrow p + e^- \tag{5.1.5}$$

At the more fundamental level, bearing in mind that the proton has a (u u d) content and the neutron comprises (u d d), we see that these scatters must involve flavour changing between u and d quarks. Thus the basic reactions are:

$$\bar{\nu}_e + u \rightarrow d + e^+ \tag{5.1.6}$$

$$\text{and } \nu_e + d \rightarrow u + e^- \tag{5.1.7}$$

The equivalent muonic reactions are obviously allowed:

$$\bar{\nu}_\mu + u \rightarrow d + \mu^+ \tag{5.1.8}$$

$$\text{and } \bar{\nu}_\mu + d \rightarrow u + \mu^- \tag{5.1.9}$$

We can see that the beta-decays and the related neutrino and antineutrino interactions with nucleons are based on the fundamental Feynman diagrams shown in Figure 5.1.

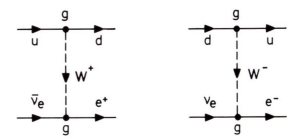

Figure 5.1. Neutrino-induced flavour changing.

If we think of the fermions as carrying a 'weak charge', g, then the coupling will be given by:

$$G_F \sim \frac{g^2/4\pi}{(q^2 + M_W^2)} \tag{5.1.10}$$

where M_W is the mass of the exchanged W particle and q is the four-momentum transfer. At momentum transfers which are small compared to the mass M_W (that is, at large distances), combining equations 5.1.10 and 5.1.1 and by analogy with the fine structure constant (see equation 3.3.2), we have

$$\frac{g^2}{4\pi} \simeq 1\cdot02 \times 10^{-5} \left(\frac{M_W}{m_P}\right)^2 \qquad (\text{c.f. } \alpha = \frac{e^2}{4\pi} \simeq \frac{1}{137}) \qquad (5.1.11)$$

(in natural units)

From the dependence of ν–p cross-sections upon momentum transfer q and energy up to 200 GeV, we can set a lower bound on the mass of the W particle. It must be heavier than 30 GeV/c^2 so that the weak force must have a range less than a few times 10^{-16} cm. ($\hbar c = 197\cdot3$ MeV fm.)

From relation 5.1.11, bearing in mind the order of magnitude of M_W, we can see that the weak charge, g, must be of the same order as the electric charge, e. In this picture, the basic weak charge is comparable with the electric charge and the weak interaction is only so feeble by virtue of the very large mass of the W appearing in the propagator (see 5.1.10). At very large momentum transfers (that is, distances less than 10^{-16} cm) the forces are of similar strength.

Let us now consider some purely leptonic weak decays. The muon decays through a weak process that involves lepton–lepton interactions:

$$\mu^- \rightarrow e^- + \bar{\nu}_e + \nu_\mu \qquad (5.1.12)$$

and the related scattering process is

$$\nu_\mu + e^- \rightarrow \mu^- + \nu_e \qquad (5.1.13)$$

(see Figure 5.2)

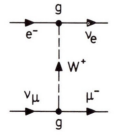

Figure 5.2. Inverse muon decay.

Similarly,

$$\bar{\nu}_\mu + e^+ \rightarrow \mu^+ + \bar{\nu}_e \qquad (5.1.14)$$

A positive pion or kaon can decay into a muon and a neutrino:

$$\pi^+ \text{ or } K^+ = (u\bar{d}) \text{ or } (u\bar{s}) \rightarrow \mu^+ + \nu_\mu \qquad (5.1.15)$$

which is obviously simply related, in the pion case, to the reaction (5.1.8). Similarly, the negative pions and kaons can decay through

$$\pi^- \text{ or } K^- = (d\bar{u}) \text{ or } (s\bar{u}) \rightarrow \mu^- + \bar{\nu}_\mu \qquad (5.1.16)$$

which again in the pion case is related to the process 5.1.9.

5.2. Gauge theories

Before we go on to see how the weak interaction can be put into a common framework with quantum electrodynamics (QED) in the combined 'electroweak' theory, we must first understand some rather technical aspects of symmetries and classes of field theory.

Let us first consider a little more about symmetries in general. We say there is a symmetry if some pattern repeats itself when a certain transformation is applied. For example, one of the best known symmetries in nature is the snowflake, which looks identical every time it is rotated through 60° in its own plane. These rotations are said to be commutative, which means that the order in which they are applied does not affect the end result. The snowflake thus has commutative symmetry under the two-dimensional 60° rotations in its plane. (The idea of commutating operations is a familiar one in ordinary algebra where, for example, the order of multiplication makes no difference to an end result. On the other hand, if we are dealing with matrix operations or differentials it is not uncommon to have non-commutative properties.)

We also speak of 'global' symmetries in physics. These occur when the relevant laws of physics remain unchanged when a particular transformation is applied everywhere in the space of interest, at the identical moment. For example, if we have an array of static electric charges, a certain electric field pattern is set up and the potential differences determine the forces that are acting between the charges. These forces would be quite unaffected by raising the potential of the whole system to some high, but arbitrary, voltage. The absolute potentials do not affect the forces, only the differences. Changing the overall potential at all points in the system at the same time is an example of global transformation. (The idea of an arbitrary origin of potential energy is familiar also in the context of ordinary mechanics.)

Within Maxwell's classical theory there is a deeper and more restrictive symmetry. As soon as electric charges are free to move, we have both magnetic and electric fields with corresponding potentials. The Maxwell equations have what is called 'local' symmetry in that a transformation can be applied which is different from point to point and yet still leaves the laws of force which govern the motions of the charges quite unchanged. This works by a mechanism in which a local change of electric potential can be locally compensated by an appropriate change in the magnetic potential, maintaining the same electric and magnetic field values at each point. Maxwell's classical theory is said to be a 'gauge' theory with local symmetry.

Let us now go on to consider what happens when relativity and quantum mechanics are introduced and we have a full field theory such as QED. Another more complicated, but related, type of gauge invariance enters the picture. When we describe a charged particle, such as an electron, in terms of a wave-function, Ψ, which may represent a wave packet, there is always an arbitrary 'phase factor' entering into the description. We cannot measure the phase of Ψ, even in principle, as all observable effects in quantum theory depend upon $|\Psi|^2$, not upon the individual components. There must therefore be a symmetry with respect to transformation which corresponds to changing this phase factor.

Mathematically, we can express this idea as follows. Let the initial wave-function and its complex conjugate be Ψ_i and Ψ_i^*, respectively. Now apply an arbitrary phase transformation $e^{i\phi}$, where ϕ is a constant at all points in space and time. We will then have a final wave function, Ψ_f, given by

$$\Psi_f = e^{i\phi}\Psi_i$$

and
$$|\Psi_f|^2 = \Psi_f^*\,\Psi_f = (e^{-i\phi}\Psi_i^*)\,(e^{i\phi}\Psi_i) = \Psi_i^*\Psi_i = |\Psi_i|^2$$

We have applied a global gauge transformation to the electron's field and the particular value of the constant ϕ corresponds to 'choosing a gauge'. As we discussed in section 2.6, all symmetries are intimately bound up with conservation laws. In this example, the global gauge symmetry of QED is connected with the conservation of electric charge, which is one of the very fundamental laws of physics.

Suppose that we go on to explore what happens when the much more restrictive local symmetry is applied to this quantum electrodynamic example. This must correspond to allowing the phase change to be a function of space and time, as opposed to being just a constant. We then have the possibility of changing the phase of the electron matter wave from point to point. How can we compensate such changes locally? In fact it can be shown that we can do just that by assuming that the electron can emit and absorb photons (massless spin-1 particles). These spin-1 fields exactly perform the trick of compensation as at each emission or absorption there has to be a local phase change. The photon must be massless if we are to maintain the infinite range of the interaction. The electromagnetic field mediated by these photons in the way described in chapter 3 ensures the gauge invariance of the electron's field.

The type of gauge symmetry exhibited in the QED theory that we have just described has another interesting property. Like the snowflake example, the phase transformations are commutative. Successive transformations, each of which changes the phase, have an end result which does not depend on the order of the sequence in which they are applied. Such symmetries in field theories are known by the technical name of 'Abelian' (after a nineteenth century Norwegian mathematician, Niels Abel, who studied such symmetry classifications). This commutation property, in the case of the electron, implies that the final phase change will not depend on the exact ordering of a sequence of emissions and absorptions of the virtual photons. In summary, we can now class QED as an 'Abelian gauge theory' with local symmetry.

In the 1950s, Chen-Ning Yang and Robert Mills developed these ideas further by considering isospin symmetry (see section 2.5) which turns out to be non-commutative, and the resulting transformation does depend upon the ordering of several transformations. (This is typical of a situation in which you consider rotations in three dimensions.) This Yang–Mills theory is thus said to be 'non-Abelian'. They looked at the consequences of forcing a local symmetry in the isospin case, more or less as a technical exercise. They found that two additional massless 'photons' were required to effect the required compensation and moreover one must carry a positive charge and the other a negative charge. Such massless but charged bosons would have a very profound effect if they existed in nature and as no such effects are seen the whole approach was quietly dropped! It seemed to be a case of a theory looking for some phenomena.

However, we will see in the next section that the electroweak theory combining the weak interactions and QED is a non-Abelian theory of the general Yang–Mills type. The old weak-interaction theory originated by Fermi was a useful phenomenological model but, amongst other things, it lacked the property of local symmetry.

5.3. *Introduction to the Weinberg–Salam model*

The main difficulty with the Yang–Mills theory (see last section) was that it could not be directly physically applicable as none of the forces of nature involve mediating particles which are massless and yet carry electric charge. In particular, they could not be associated with the W^+ and W^- intermediate vector bosons required by the charged current weak interaction data discussed in section 5.1. The reason that the masses came out to be exactly zero in that theory is that the local symmetry was exact. We have seen before that an approximate or 'broken' symmetry can generate different masses. The problem is, how can the symmetry be 'spontaneously' broken and give masses to the two massless charged 'photons' to construct a physically useful theory? The general idea of how a perfect symmetry can be spontaneously broken is perhaps made clearer by simple examples. A perfectly symmetric roulette wheel is spun and the ball allowed to spin around its rim in the opposite direction. Here we have perfect symmetry as long as the relative motion is maintained. Once the ball comes to rest in a particular groove the symmetry is broken. Again, consider a rigid pendulum precariously pointing exactly upright (in what is called metastable equilibrium). A tiny perturbation can cause it to fall in an anticlockwise or clockwise direction and as soon as the fall begins the symmetry is destroyed. Another often-quoted example is that of a round dinner table set with napkins symmetrically placed at each setting. As soon as the first diner picks up his napkin on, say, his left-hand side the symmetry is broken.

What has all this got to with weak interactions? In the Yang–Mills theory, the perfect symmetry is between the three massless 'photons' with $-1, 0$ and $+1$ units of charge. Peter Higgs suggested a mechanism which could effect the required symmetry breaking and give the two charged 'photons' some mass. These extra 'photons' have spin-1 and one might think that they should have three spin states (as in Appendix B.4). In this special case of free massless objects travelling at the speed of light, one degree of freedom is lost and there are actually only two allowed spin states according to the normal rules of quantum mechanics†. If you force these particles to acquire mass, then they will need one extra spin state each. In quantum mechanics we have to account for these suddenly acquired quantum states. Higgs suggested that if there are extra spin-0 (scalar) bosons (usually called Higgs particles) then some of these could disappear, giving up their spin state to the charged 'photons' and allowing them to acquire the necessary mass.

After some early work by Sheldon Glashow, Steven Weinberg, Abdus Salam and John Ward worked on the weak interaction theory and showed that the imposition of local symmetry on the invariance of the interaction gives rise in this case to four massless 'photons', the W^+, W^-, W^0 and B^0. (The three W's come from the SU(2) symmetry of the weak isospin and the B^0 comes from the U(1) symmetry of the electromagnetic theory. The combined electroweak theory then corresponds to an overall $SU(2) \times U(1)$ theory.) The physical mediators of the forces are then the W^+ and the W^- for the charged weak currents and two orthogonal superpositions of the W^0 and the B^0 giving the Z^0 for the weak neutral currents and the photon for the electromagnetic force. The coupling strength associated with the W^+, W^- and the W^0 is called g_2 and the B^0 has a coupling g_1.

The Higgs mechanism was then used to break the symmetry to allow three of these particles to acquire a spin state and hence mass by 'eating up' (as Salam put it) three Higgs particles. However, the detailed theory requires four Higgs particles to be generated and this leaves one massless particle (the photon itself) and one real Higgs particle. The three Higgs fields are thus used up to allow for the W^+ and W^- and, slightly surprisingly, a massive neutral boson, the Z^0. The 'uneaten' Higgs particle remains, as does the normal massless neutral photon. The three massive mediators of the weak interaction and the single massless mediator of the electromagnetic interaction are thus tied together in one theory.

---

† Note that a virtual photon has an effective mass and therefore when mediating the electric force between two charges, it has three sub-states, giving the vector nature of the interaction.

The final requirement was to show that the resulting 'electroweak' theory could be used to calculate cross-sections and the usual field theory problem of divergent self energies arose, as in the QED example of Figure 3.5. The renormalization problem was solved in 1971 by Gerard 't Hooft, Martin Veltman and Benjamin Lee and is now on a rather firm footing.

Let us try to summarize these rather difficult concepts. The Weinberg–Salam theory embodies the weak and the electromagnetic forces in a single 'electroweak' theory. The hypothesis of local symmetry with respect to the leptonic equivalent of isospin yields four massless Yang–Mills particles which are photon-like but have charges $+1$, -1, 0 and 0. The Higgs mechanism of spontaneous symmetry breaking provides four Higgs fields, three of which are eaten up by the Yang–Mills particles giving the massive W^+, W^- and Z^0. We are left with the fourth Yang–Mills particle as the normal massless uncharged photon and one spin-0, probably massive, Higgs particle is also left over. (We should therefore add at least one Higgs particle to our list of fundamental constituents in the standard model as given in Table 4.9).

The 'Weinberg angle', Θ_w, is a measure of the mixing of the W^0 and B^0 to give the photon and the Z^0. We then have a ratio of electric to weak charges given by

$$\sin \Theta_w = e/g \qquad (5.3.1)$$

In the simplest version of the theory, the masses of the W^0 and the Z are then related by

$$M_w^2 = M_z^2 \times \cos^2 \Theta_w \qquad (5.3.2)$$

The only free parameter (which can be fixed by comparison with low energy data) is the Weinberg angle. In fact $\sin^2 \Theta_w$ is found to be about $0 \cdot 23$ and the masses of the particles are then predicted to be

$$\begin{aligned} M_w &= 77 \cdot 8 \, \text{GeV}/c^2 \\ M_z &= 88 \cdot 7 \, \text{GeV}/c^2 \end{aligned} \qquad (5.3.3)$$

After many minor corrections are taken into account the current theory estimates values as follows:

$$\begin{aligned} M_w &= 83 \cdot 0 \, \text{GeV}/c^2 \\ M_z &= 93 \cdot 8 \, \text{GeV}/c^2 \end{aligned} \qquad (5.3.4)$$

(with theoretical uncertainties of around 2 to 3 GeV/c^2).

The experimental value of $\sin \Theta_w$ determines $e/g \approx 0 \cdot 23$, which is in agreement with the argument in section 5.1 which suggested that the two charges would be comparable in magnitude, the large mass of the W (and Z) being responsible for the feeble nature of the weak force.

The other striking prediction of the thory is the existence of a massive neutral mediator of the weak force, the Z^0. This implies that so-called 'neutral current' events should be seen in the weak interactions. For example, in muonic neutrino scattering off nucleons there should be events with no charged leptons in the final state. The fundamental reactions would then be

$$\nu_\mu + d \to \nu_\mu + d \tag{5.3.5}$$

$$\text{and} \quad \nu_\mu + u \to \nu_\mu + u \tag{5.3.6}$$

to be compared with the charged current reactions 5.1.8 and 5.1.9. We should be able to observe

$$\nu_\mu + \text{nucleon} \to \nu_\mu + \text{nucleon} \tag{5.3.7}$$

at low energies, and as the energy increases we will see:

$$\nu_\mu + \text{nucleon} \to \nu_\mu + \text{several hadrons} \tag{5.3.8}$$

These processes are depicted in Figure 5.3. We would also expect neutral current events in lepton–lepton scattering, for example:

$$\bar{\nu}_\mu + e^- \to \bar{\nu}_\mu + e^- \tag{5.3.9}$$

Thus the electroweak theory predicts the masses of the W^+, W^- and the Z^0 and the existence of neutral current weak interactions.

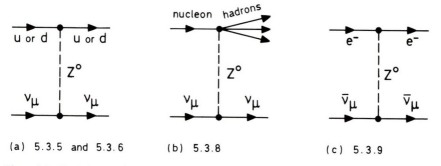

(a) 5.3.5 and 5.3.6   (b) 5.3.8   (c) 5.3.9

Figure 5.3. Neutral current processes.

5.4. Experimental tests for the electroweak theory

The first real confirmation of the validity of the electroweak theory came in the summer of 1973 with the discovery of neutral currents in an experiment at

CERN, Geneva. A large number of neutrino interactions were observed in a very big bubble chamber (called 'Gargamelle') filled with heavy liquid (liquid Freon). A proportion of these appeared to have no charged leptons among the produced secondary particles which emerged from the interactions. These events were quickly interpreted as examples of neutral current events of the type 5.3.8:

$$\nu_\mu + (\text{a nucleon in the nucleus}) \rightarrow \nu_\mu + \text{several hadrons}$$

in which no quark flavour-changing process occurs, and are associated with the exchange of the neutral massive Z^0 boson. We will now look in a little more detail at how this evidence was produced.

The CERN proton synchrotron (PS) was used to accelerate proton beams up to 26 GeV. The beam was extracted and fired into a heavy metallic target, where strong interactions occurred producing many secondary hadrons, mostly pions and some kaons. Positively charged particles were then separated magnetically and the forward-going particles were allowed to pass through an empty 'drift' region where some of the mesons decayed. The charged pions decay almost entirely by the mode

$$\pi^+ \rightarrow \mu^+ + \nu_\mu \tag{5.4.1}$$

The decays of the charged kaons are more complicated. The predominant decay (63%) is similar to that of charged pions, namely

$$K^+ \rightarrow \mu^+ + \nu_\mu \tag{5.4.2}$$

However, $4 \cdot 8\%$ decay via the process

$$K^+ \rightarrow \pi^0 + e^+ + \nu_e \tag{5.4.3}$$

which means that some electronic neutrinos are also produced among the decay products. Thus, by selecting positively charged hadrons, a fairly pure sample of muonic neutrinos could be produced and, similarly, muonic antineutrinos could be obtained from a negative hadron beam.

All the particles were then allowed to impinge on a huge 'dump' consisting of iron and concrete whose function was to remove the hadrons and the penetrating muons. The hadrons were completely removed in the first few metres of the shielding through a cascade of strong interactions and even the relatively penetrating muons eventually lost their energy through ionization processes and were brought to rest.

At the far side of the dump, only the extremely penetrating neutrinos or antineutrinos survived and a broad beam of such particles was allowed to enter the huge bubble chamber. The heavy liquid provided many nuclear targets for the very rare neutrino collisions. After photographing, measuring and reconstructing, the events were seen to consist mostly of neutrino interactions in which there was a charged lepton amongst the secondary particles and which corresponded to the charged current process

$$\nu_\mu + \text{nucleon} \rightarrow \mu^- + \text{hadrons} \qquad (5.4.4)$$

However, in about one quarter of the cases, there was no charged muon visible

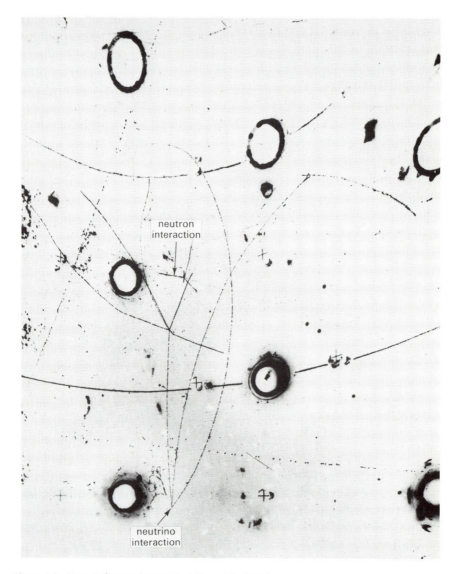

Figure 5.4. A neutral current event ($\nu_\mu + \text{N} \rightarrow \nu_\mu + \text{hadrons}$).

(see, for example, Figure 5.4), and these events seemed to be examples of the process 5.3.8 discussed above, that is,

$$\nu_\mu + \text{nucleon} \rightarrow \nu_\mu + \text{hadrons}$$

Before claiming evidence for neutral currents, the team of physicists carefully checked for possible background processes that could simulate the events. The most likely source of confusion arose from the possibility of neutron interactions with a similar topology to the neutrino examples. Like the neutrino, the neutron is uncharged and therefore leaves no track in the bubble chamber, but neutrons interact strongly with other nucleons in the liquid and usually do not produce a charged lepton amongst the secondaries. The neutrons arise from the fact that some neutrinos interact in the huge dump of material and forward-going produced neutrons can enter the bubble chamber, where they then interact. In Figure 5.4, there happens to be a neutron from the neutrino interaction in the bubble chamber, which then interacts. A study of such events enabled an upper limit to be placed on this background at the 4% level of all 'normal' (charged current) interactions. This compares to the neutral current event which were appearing at the rate of 23% of the charged current events. Furthermore, the distribution of the neutral current events along the length of the bubble chamber liquid traversed was more or less flat, whereas it would have peaked at the front of the chamber if they corresponded to strong interactions of neutrons produced upstream.

Similar neutral current events were also obtained from antineutrino beams and a small number of pure leptonic neutral current events were seen.

$$\bar{\nu}_\mu + e^- \rightarrow \bar{\nu}_\mu + e^- \tag{5.4.5}$$

These are relatively rare events as the electrons form a very small fraction of the total mass of the heavy liquid target, but they were nevertheless very interesting examples of Z^0 exchange processes.

Very quickly after the CERN discovery, other laboratories reported similar evidence for neutral currents using a variety of electronic detectors as well as bubble chambers.

A second very delicate experiment showed that the Z^0 exists and demonstrated the mixing of weak and electromagnetic interactions in a very direct way. In electron–deuteron scattering experiments at Stanford, California, small parity-violating effects were observed in which the interaction rate for left-handed electrons was seen to be about 0·01% more frequent than right-handed electron collisions. The interpretation is that the normal parity-conserving electromagnetic scattering mode with a photon exchange (Figure 5.5(*a*)) is mixed with the virtual Z^0 exchange mechanism (Figure 5.5(*b*)) and this second process can violate parity conservation.

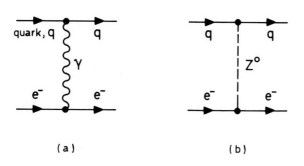

Figure 5.5. Electroweak mixing.

There are some indications of similar, very small, parity-violating effects in atomic physics. The existence of the weak neutral force acting between the positively charged nucleus and the electrons in the atom can again give rise to mixing with the normal photon coupling of the electromagnetic forces. However, for the time being the results of these experiments are not completely clear as very precise atomic calculations of complex atoms are also required to untangle the experimental observations.

The final, and even more remarkable, confirmation of the predictions of the Weinberg–Salam model came in 1983 when the W and Z particles were directly observed and their masses were found to be amazingly close to their expected values.

Why had it taken so long to see these particles when all their expected properties were well known? The answer is simply that the available accelerators and colliders could not produce a high enough available energy. For example, the 400 GeV SPS machine at CERN gives only about 20 GeV in the centre-of-mass and the highest energy beams in the intersecting storage rings at CERN were 32 GeV protons, giving about 60 GeV of available energy. The electron–positron collider at DESY provides a maximum of around 40 GeV. In all these cases, the available energy is well below the masses of the W and Z particles which are in the 80 to 90 GeV/c^2 region. In fact the problem is even worse than this, because if you have a proton beam of, say, 30 GeV/c, then about half the momentum is carried by the gluons and the remaining momentum is shared between the u, u and d quarks within the proton. Thus, typically each quark will carry only about 5 GeV/c and in a colliding-beam machine with two such beams moving in opposite directions the effective available energy in quark–quark collisions is nearer to 10 GeV than the 60 GeV you might have expected. The other point to bear in mind in designing the experiment is that quark–antiquark collisions can be more suitable as the total annihilation energy is then available for particle production and, moreover, both beams of protons and antiprotons can be stored in the same magnetic ring in a collider as the electric charges are equal in magnitude but opposite in sign.

In a very bold plan, largely inspired by Carlo Rubbia, it was decided to construct a system at CERN which would give the very highest possible energy quark–antiquark collisions by turning the SPS (400 GeV) accelerator into a proton–antiproton storage ring collider device. For technical reasons it was realized that the highest energy particles that could be stored in such a system would be 270 GeV, giving a total available energy of 540 GeV and even then the typical available energy in each quark–antiquark collision would only be about 90 GeV. This meant that W's with an expected mass of about 80 GeV/c^2 would be produced more copiously than the Z's with a mass in the region of 90 GeV/c^2.

The main problem with this proposal was to find a method of producing sufficient numbers of antiprotons to give a reasonable chance of interactions in the collider. Protons were first accelerated in the 'small' CERN proton synchrotron, yielding 26 GeV protons which could be ejected and fired at a heavy target. Antiprotons are rather infrequently produced in these collisions and those with an energy of about 3·5 GeV were magnetically selected and stored in a specially constructed small storage ring called the antiproton accumulator (or 'AA', see Figure 5.6). This process is repeated many times until a sufficiently high-intensity antiproton beam is accumulated in the AA. In order for the accumulation process to work, a new principle was developed at CERN by Simon Van der Meer called 'stochastic cooling'. In order to be able to pack a large number of antiprotons into the stored beam, the random motions of the particles must be reduced. This is done by having pick-up electrodes somewhere in the circumference of the ring, which sense deviations of the particles. A signal is then sent across the diameter of the ring to arrive before the particles and a compensating 'kick' is there applied so that on average a convergence is produced.

This subtle method lies at the core of the success of the whole project. Having obtained a large flux of antiprotons in this manner, the beam is sent to the SPS where the particles are accelerated up to 270 GeV. Protons can then be injected and accelerated in the normal way up to the same energy and will travel in the opposite direction in the same vacuum pipe (as the two beams have opposite electric charges). The protons and antiprotons are not spread all around the circumference but are deliberately 'bunched' in such a way that they only collide at certain preselected points around the beam pipe and at these positions experiments can be mounted.

Two particular experiments were aimed specifically at discovering the W and Z particles. Experiment UA1 ('underground area 1') was led by Carlo Rubbia and UA2 was led by Pierre Darriulat. A photograph of the experiment UA1 is shown in Figure 1.4. The UA1 detector was designed to cover a solid angle of nearly 4π steradians to detect as many of the products of p p̄ interactions as possible. This huge detector ($10 \times 6 \times 6\,m^3$) measures the momenta or energy of the individual particles and identifies electrons, muons, photons and hadrons. Searches were made for the production of W's and their subsequent decays into one of the modes.

Figure 5.6. The CERN Antiproton Accumulator (AA).

$$W^+ \rightarrow e^+ + \nu_e \qquad (5.4.6)$$
$$W^- \rightarrow e^- + \bar{\nu}_e \qquad (5.4.7)$$
$$W^+ \rightarrow \mu^+ + \nu_\mu \qquad (5.4.8)$$
$$\text{and } W^- \rightarrow \mu^- + \bar{\nu}_\mu \qquad (5.4.9)$$

Early in 1983 the first tentative results were reported and a handful of events of this kind, with single, isolated, charged leptons of very high energies at wide angles were seen in both experiments UA1 and UA2. During the following year more than 50 events were reported from UA1 of the positron and electron types (5.4.6 and 5.4.7) and a few muonic events were also seen. (These muonic events were technically more difficult to detect and there was a more limited acceptance for muon identification, but after suitable corrections are made the expected

number of these events was indeed observed.) The W's can also decay hadron-
ically as

$$W^{\pm} \rightarrow q + \bar{q} \qquad (5.4.10)$$
$$\quad \ \ \big\lfloor \ \ \overset{\hookrightarrow}{} \text{ hadron jet}$$
$$\qquad \overset{\hookrightarrow}{} \text{ hadron jet}$$

In June 1983 experiments UA1 and UA2 both reported the first sightings of
the production of Z^0 particles, having obtained four events each which fitted one
of the hypotheses

$$Z^0 \rightarrow e^+ + e^- \qquad (5.4.11)$$

$$\text{or } Z^0 \rightarrow \mu^+ + \mu^- \qquad (5.4.12)$$

Most of the observed events were again electronic rather than muonic for the
same reason as in the W search.

A typical Z^0 event is shown in Figure 5.7, where a large number of particles
are produced but a very high-energy electron and positron emerge at large angles

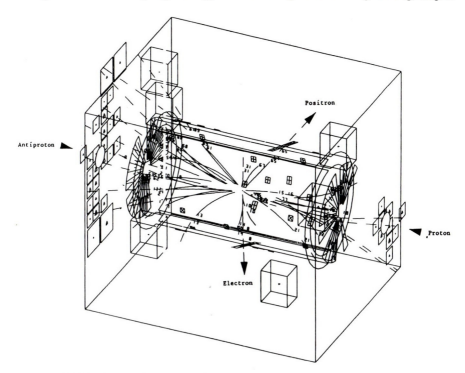

Figure 5.7. Z^0 production in a p–p̄ collision with 540 GeV available energy. The Z decays to an
ee pair, both particles having a very large transverse energy.

to the axis of the colliding proton and antiproton, the corresponding mass being just right for the hypothesis of Z^0 production followed by a two-particle decay. There are essentially no competing background events to this process. As in the case of the W's, since the initial discovery many examples of Z^0 decay, into both decay channels, have been seen.

The current (June 1984) best values for the masses of the $W^\pm$ and the Z^0 and the theoretical expectations are shown in Table 5.1.

Table 5.1. Masses of the $W^\pm$ and Z^0 particles (in GeV/c^2).

	Experimental values (value ± statistical ± systematic errors)		Theoretical values (with theoretical 'uncertainties')
	UA1	UA2	
M_w	$80 \cdot 9 \pm 1 \cdot 5 \pm 2 \cdot 4$	$83 \cdot 1 \pm 1 \cdot 9 \pm 1 \cdot 3$	$83 \cdot 0 \pm 3 \cdot 0$
M_z	$95 \cdot 6 \pm 1 \cdot 5 \pm 2 \cdot 9$	$92 \cdot 7 \pm 1 \cdot 7 \pm 1 \cdot 4$	$93 \cdot 8 \pm 2 \cdot 5$

It is very clear that the agreement between the predicted and measured values of these masses represents an enormous triumph for the electroweak theory. From the measured masses, the Weinberg angle can be extracted using the relation 5.3.2. For example, the UA1 result gives

$$\sin^2\Theta_w = 0 \cdot 226 \pm 0 \cdot 008 \pm 0 \cdot 014 \qquad (5.4.13)$$

For their parts in making the discovery possible, Rubbia and Van der Meer were awarded the 1984 Nobel Prize for physics.

5.5. Quark flavour-changing

The weak force provides the only interactions that are capable of changing the flavours of quarks. For this reason the electroweak theory is often called quantum flavour dynamics (or 'QFD').

Information on the rules for flavour-changing has come mainly from studying the weak decays of hadrons into either pure hadronic or mixed hadronic and leptonic final states. By determining the rates for various observed transitions and noting which quark changes must be occurring, some empirical rules have been developed.

First, in charged current events, transitions involving d to u interchanges go about 20 times faster than decays involving s to u transitions. For example, normal beta-decay processes involve d to u transitions, that is,

$$\begin{aligned} n &\to p &+ e^- + \bar{\nu}_e \\ \text{or } (u\,d\,d) &\to (u\,u\,d) &+ e^- + \bar{\nu}_e \end{aligned} \qquad (5.5.1)$$

Now consider the 'semi-leptonic' decay of the Σ^- which involves an s to u transition, that is,

$$\Sigma^- \to \quad n \quad + e^- + \bar{\nu}_e$$
$$\text{or } (d\,d\,s) \to (d\,d\,u) \; + e^- + \bar{\nu}_e \tag{5.5.2}$$

After taking out kinematic factors, what is seen is that the sum of the amplitudes squared for processes like 5.5.1 and 5.5.2 is equal to the square of the amplitude for the basic leptonic weak interactions such as

$$\mu^- \to e^- + \bar{\nu}_e + \bar{\nu}_\mu \tag{5.5.3}$$

Nicola Cabibbo suggested that the explanation could be that the basic weak couplings involved are as in Table 5.2.

Table 5.2. Cabibbo couplings.

Interaction	Coupling	Transition rate
ν_e–e	g	g^2
d–u	$g \cos \Theta_c$	$g^2 \cos^2 \Theta_c$
s–u	$g \sin \Theta_c$	$g^2 \sin^2 \Theta_c$

Θ_c is the mixing angle or Cabibbo angle and g is the weak charge.

Such a scheme will certainly account for the relationship between the observed decay rates. If the individual rates are put in, we find that Θ_c is about $0\cdot23$ radians. Cabibbo suggested that one should think of the d and s quarks taking part in the weak interactions as a linear superposition of states given by

$$d_c = d \cos \Theta_c + s \sin \Theta_c \tag{5.5.4}$$

so that we should think of the first generation doublets in the weak interaction as

$$\begin{pmatrix} \nu_e \\ e^- \end{pmatrix} \text{and} \begin{pmatrix} u \\ d_c \end{pmatrix} = \begin{pmatrix} u \\ d \cos \Theta_c + s \sin \Theta_c \end{pmatrix} \tag{5.5.5}$$

The second main experimentally observed empirical rule was that whenever a decay occurs through the neutral current process, the number of strange quarks is unchanged before and after the decay. For example, in the decay of the positive kaon we see many charged current decays in which there are no strange particles in the final state, but there is only a very low upper limit for the occurrence of equivalent decays through neutral currents.

$$\frac{K^+ \to \pi^+ + \nu + \bar{\nu}}{K^+ \to \pi^0 + \mu^+ + \nu_\mu} \text{ is less than } 10^{-5} \tag{5.5.6}$$

Sheldon Glashow, John Iliopoulos and Luciano Maiani invented a model (now always known as the 'GIM' mechanism after the initials of the authors of the famour paper) which accounted for this observation. They suggested that a second weak quark doublet should consist of

$$\begin{pmatrix} c \\ s_c \end{pmatrix} = \begin{pmatrix} c \\ s\cos\Theta_c - d\sin\Theta_c \end{pmatrix} \tag{5.5.7}$$

where c is the fourth (charm) quark, which at the time of the suggestion had not yet been discovered. The combination of s and d quarks that appears here, together with the combination that couples to the u quark, gives just the required cancellations. In this model there will be four basic diagrams describing the weak neutral couplings of the quarks as in Figure 5.8.

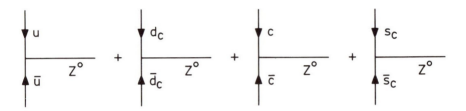

Figure 5.8. The 'GIM' mechanism contribution

Algebraically we then have the following couplings for the quark–Z vertex corresponding to the four diagrams:

$$u\bar{u} + d_c\bar{d}_c + c\bar{c} + s_c\bar{s}_c$$

$$= [u\bar{u} + [(d\cos\Theta_c + s\sin\Theta_c).(\bar{d}\cos\Theta_c + \bar{s}\sin\Theta_c)] + c\bar{c}]$$
$$+ [(s\cos\Theta_c - d\sin\Theta_c).(\bar{s}\cos\Theta_c - \bar{d}\sin\Theta_c)]$$

$$= [u\bar{u} + (d\bar{d} + s\bar{s})\cos^2\Theta_c + (s\bar{s} + d\bar{d})\sin^2\Theta_c + c\bar{c}]$$
$$+ [(s\bar{d} + \bar{s}d - \bar{s}d - \bar{d}s).\sin\Theta_c.\cos\Theta_c]$$

The first main bracket gives the term where no strangeness-changing occurs, and all the terms in the second bracket cancel. We can therefore account for the observation that there are no strangeness-changing neutral currents. Strictly, this cancellation only works perfectly if the mass of the charmed quark is equal to the mass of the u quark. However, just the degree of allowed higher-order non-cancellations can occur (corresponding to the observed upper limits) if the mass of the c quark is around 1 to $3\,\mathrm{GeV}/c^2$. We now know that the mass is around $1\cdot5\,\mathrm{GeV}/c^2$ and so the model fits the data very well.

The argument can be extended to account for the third doublet of quarks that involves the top and bottom quarks. The most general weak mixing for the six constituent quarks of the standard model is described by the 'Kobayashi–Maskawa' matrix, which involves three mixing angles (and an additional phase angle which could be related to the small T violation (see p. 25) seen in neutral kaon decay). The Cabibbo angle is now just one of these three mixing angles in this more general model.

The present data, although by no means complete, suggest that all three angles are small, which implies that the most likely decays form part of the chain

$$t \rightarrow b \rightarrow c \rightarrow s \rightarrow u \qquad (5.5.7)$$

and skipping a generation (e.g. $b \rightarrow u$) is not favoured (see Figure 5.9).

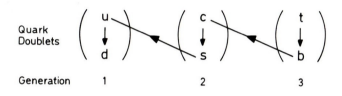

Quark Doublets

Generation   1   2   3

Figure 5.9. The most favoured weak decay chain.

CHAPTER 6

the strong interaction

6.1. Coloured quarks revisited

Within the framework of the standard model, quantum chromodynamics, 'QCD' as it is often called, is at least a good candidate theory of the strong force. This theory is not on the same sure footing as quantum electrodynamics (QED), nor is it even as well tested as the electroweak theory described in the previous chapter. However, QCD is the best theory so far suggested and it has had many successes in describing some of the most basic aspects of the strong force. It is in some ways simpler than the electrweak theory because the basic symmetry involved is exact.

One of the main ingredients of the theory is the idea that quarks carry a new kind of charge which we call 'colour charge' or, more briefly, just 'colour', already mentioned in section 4.6. In electrostatics we are used to the idea that we just have positive and negative charges: neutral systems are produced when we add one equal negative charge to each positive charge. In QCD, the colour charge exists in three varieties which we shall call red (R), green (G) and blue (B). Of course it must be realized that quarks are not coloured in the normal sense of the word, but rather that the colour charge has similar additive properties to those of real colour. The antiquarks are assumed to come in any of the three corresponding anticolours ($\bar{R}$, $\bar{G}$ and $\bar{B}$). It is also assumed that a colour charge and its anticolour charge attract (just like opposite electric charges), but in addition three unlike colours also attract whereas two unlike colours repel each other. The colour symmetry is exact so that the R, G and B quarks of a given flavour all have the same mass. Successive colour transformations will cause the colours to change into each other cyclically.

All hadrons that we have so far observed are neutral in their colour charge. We say that they are 'colourless' or 'white' and such states could be constituted in various ways. The three quarks in a baryon must be selected in such a way that one is R, one is G and one is B, irrespective of the flavours of the quarks. Similarly, an antibaryon must have one of each of the anticoloured antiquarks (one $\bar{R}$, one $\bar{G}$ and one $\bar{B}$). The mesons (and antimesons) must have a quark of a particular colour and an antiquark with the anticolour of the quark. Some examples are given in Table 6.1. At all times the hadron must be colourless. Even when quark–quark interactions occur within the hadron through gluon exchanges (see next section), the resulting colour changes must happen in such a way that the whole hadron remains in its colourless state.

In terms of the colour symmetry property, we can understand what is happening in the following way. A global colour transformation (see section 5.2 for the meaning of 'global') will change all the quark colours simultaneously, keeping the overall colour 'white'.

What is the evidence that such colour charges exist? First, we have already mentioned (section 4.6) that there is a problem in the existence of baryons constructed from three identical quarks such as the Δ^- (d d d), the Δ^{++} (u u u) and the Ω^- (s s s). The Pauli exclusion principle for the fermionic quarks requires at least one quantum number to be different for each of the three quarks in such systems. The idea of colour charge immediately overcomes this problem.

Table 6.1. Examples of hadronic colour charges.

Particle		Quark flavour and colour		
Proton, p		u	u	d
		R	G	B
	or	R	B	G
	or	B	G	R
Antiproton, $\bar{\text{p}}$		$\bar{\text{u}}$	$\bar{\text{u}}$	$\bar{\text{d}}$
		$\bar{\text{R}}$	$\bar{\text{G}}$	$\bar{\text{B}}$
	or	$\bar{\text{R}}$	$\bar{\text{B}}$	$\bar{\text{G}}$
	or	$\bar{\text{B}}$	$\bar{\text{G}}$	$\bar{\text{R}}$
Positive pion, π^+			u	$\bar{\text{d}}$
			R	$\bar{\text{R}}$
	or		G	$\bar{\text{G}}$
	or		B	$\bar{\text{B}}$
Omega minus, Ω^-		s	s	s
		R	G	B

There are other cogent reasons for believing that quarks carry this extra colour degree of freedom. For example, let us consider the decay of the neutral pion in two γ-rays. It is assumed that this process occurs as in the Feynman diagram shown in Figure 6.1. This is an electromagnetic decay process and, as we discussed in section 3.3, depends on the coupling of the electric charges of the quarks to the two photons. The probability for this decay will also depend upon how many varieties of the q$\bar{\text{q}}$ pairs contribute and in fact the width (see section 2.2) of the decay is proportional to the square of this number of quark types. A detailed calculation is in agreement with the measured value only if we assume that there are three times as many quark types as there are flavours. This is just what would be predicted if there are three colour attributes in addition to the various flavours.

Similar tests have been carried out in the study of the products of high-energy electron–positron collisions in storage machines, like that at Frascati near

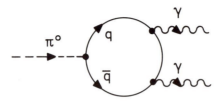

Figure 6.1. Decay of the neutral pion.

Rome, the PETRA collider at the DESY laboratory in Hamburg and that at Stanford, California. Experimentally, we can compare the cross-section for the production of hadrons with that for the production of a pair of leptons such as μ^- and μ^+. These processes are depicted in Figure 6.2 and fundamentally depend on the probability for a virtual photon to produce a $q\bar{q}$ pair or a $\mu^- \mu^+$ pair.

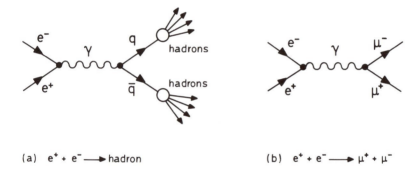

(a) $e^+ + e^- \longrightarrow$ hadron   (b) $e^+ + e^- \longrightarrow \mu^+ + \mu^-$

Figure 6.2. Production of (a) hadrons and (b) $\mu^+\mu^-$ pairs in e–e collisions.

If we measure the ratio of cross-sections, R, defined by

$$R = \frac{\sigma(e^+ e^- \to \text{hadrons})}{\sigma(e^+ e^- \to \mu^+ \mu^-)} \qquad (6.1.1)$$

at a cms energy of about 2 to 3 GeV, which is below threshold for the production of the ψ (the lowest $c\bar{c}$ state), we find a value of about $2 \cdot 0$. Above 10 GeV, the measured value is found to be about 11/3. Are these values what we would expect? Theoretically we expect the value of R to vary as the sum of the squares of the charges of all types of quarks that play a part in the process shown in Figure 6.2(a). Thus, we expect in the 2–3 GeV region that we will have contributions from u, d and s quarks only, each having the possibility of carrying three different colour charges (R, G and B). Thus,

$$R \approx 3 \times [(+2/3)^2 + (-1/3)^2 + (-1/3)^2] = 2$$

colour   u   d   s

factor

Above 10 GeV, we also have the possibility of producing charm and beauty pairs and we therefore expect

$$R \approx 3 \times [(+2/3)^2 + (-1/3)^2 + (-1/3)^2 + (+2/3)^2 + (-1/3)^2] = 11/3$$

colour   u   d   s   c   b

factor

The remarkable agreement between theory and experiment gives us a good deal of confidence in the basic ideas of colour, flavour and fractional charges.

We have said that the colour symmetry exhibited by the quarks is 'exact' in the sense that quarks of the same flavour but carrying different colour charges all have exactly the same mass. (This is in marked contrast to the flavour symmetry seen in the baryon and meson multiplets where members of a given set, for example, the spin 3/2 baryon decuplet, all have different masses. In that sense the flavour symmetry is only very approximate. Within the large multiplet, the isospin symmetry is also partly 'broken' as even the various charge states with the same isospin, such as the Δ^-, Δ^0, Δ^+ and the Δ^{++}, all have slightly different masses.)

6.2. Coloured gluons

We have already noted that the exchanged field particles of the strong interaction are the gluons which form an octet of bosons which carry the colour force. They not only transmit the colour force, but they themselves also carry colour charges.

The gluons are bosons with a spin quantum number of 1 and they are electrically neutral. They are also assumed to be massless. The gluons are flavourless and therefore the flavour of a quark is not changed when it emits or absorbs a gluon (unlike the absorption of the W's in the electroweak interaction).

The eight gluons are superpositions of various combinations of colour charge as shown in Table 6.2. Illustrations of the effects of quarks emitting and absorbing gluons have already been given in Figure 4.9.

Table 6.2. Gluon colour combinations.

1 $R\bar{G}$	2 $G\bar{B}$	3 $B\bar{R}$
4 $R\bar{B}$	5 $G\bar{R}$	6 $B\bar{G}$
	7 $(R\bar{R} + G\bar{G} - 2B\bar{B})\ \sqrt{6}$	
	8 $(R\bar{R} - G\bar{G})\ \sqrt{2}$	

The eight coloured gluons arise from applying the restrictive local colour symmetry. (For a discussion of what is meant by a 'local' symmetry, see section 5.2.) These eight gauge particles are required to ensure that a single quark can change its colour and locally compensate for the effect by the emission or absorption of a corresponding gluon. (This is just the kind of argument that we have seen in QED by which a local phase change transformation applied to an electron is compensated by the emission or absorption of a photon.) Figure 6.3 shows this process going on within a proton.

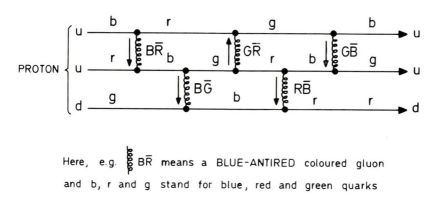

Here, e.g. $\overset{\text{BR}}{}$ means a BLUE-ANTIRED coloured gluon

and b, r and g stand for blue, red and green quarks

Figure 6.3. Gluon exchanges between coloured quarks confined in a proton.

6.3. Quantum chromodynamics (QCD)

We have said that the basic quark–quark strong interaction is assumed to take place through the exchange of a coloured gluon, just as in the electromagnetic case the interaction occurs through the exchange of a photon. Although there are superficial similarities between the strong force theory (QCD) and that of the electromagnetic force (QED) we will see that there are also important differences.

In QED the photon can only couple to an electric charge and there is therefore only one basic vertex to understand. In QCD, the gluon couples to colour charges on quarks or on other coloured gluons (see Figure 6.4). Just as in the QED and electroweak cases, QCD is a gauge theory (of the Yang–Mills type) and as the colour symmetry is exact, the strong force is gauge invariant. The strong colour charges are conserved and the masses of the eight gluons are all exactly zero.

In principle, as the gluons can couple to each other there should be bound states of gluons, called 'glueballs', and two simple examples are shown in Figure 6.5.

Another possibility would be the existence of 'hybrid' hadronic states which

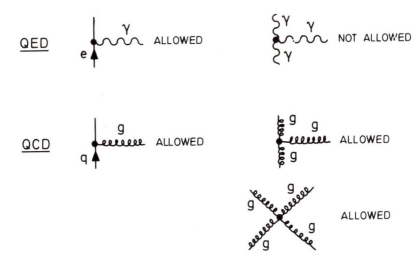

Figure 6.4. Basic QED and QCD vertices.

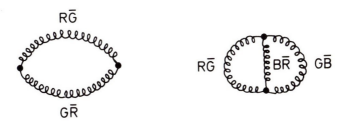

Figure 6.5. Simple examples of glueballs.

contain a mixture of quarks and gluons such that the overall hadron is still kept colourless. Such a state could be, for example, a system of one red quark, one antiblue antiquark and a B R̄ gluon. So far there is no compelling evidence for clear examples of either glueballs or hybrid hadrons and this is a little worrying from the point of view of the QCD theory.

It is clear in the case of photons that, as they are uncharged, they carry no electrical quantum number information and only electric charges can be sources of the electromagnetic field. In contrast, in QCD the field particles (gluons) carry the colour quantum number information and are therefore themselves sources of the strong field and this allows them to couple directly to other gluons.

The properties of the colour transformations are such that the order of a given sequence of transformations affects the final result. That is, the symmetry is non-commutative (like three-dimensional rotations) or non-Abelian (see section 5.2). It is this property that leads to the idea that the field bosons (gluons) are also sources of the colour field and thus carry a colour charge.

In order to understand the nature of the strong coupling it is useful to discuss again the electromagnetic coupling, this time in a little more detail. In QED we have already seen a picture (Figure 3.5(*a*)) of the free electron as a 'bare' charge surrounded by virtual photons and virtual particle–antiparticle pairs. These pairs which have in a sense 'popped out of the vacuum' have the effect of shielding the bare charge. This effect is known as 'vacuum polarization', as the charged pair of particles (usually an electron–positron pair as this is the lowest mass possibility) will be polarized by the bare charge and hence they will shield the bare electron as viewed from some distance away.

This idea of polarization shielding is more familiar in the context of an electrostatic charge in a dielectric medium. Consider a positive charge *q*, immersed in a dielectric liquid medium as in Figure 6.6. The molecules of the dielectric will be assumed to be free to align themselves as shown in the diagram. If we now bring a very small test charge into the dielectric at distances large compared to the size of the dielectric molecules, we know from simple electrostatic theory that the test charge will experience a force which is as if the positive charge *q* has been reduced by a factor $1/K$ (where K is the dielectric constant). However, at very small distances, less than the size of the dielectric molecule, our test charge sees the full positive charge *q*.

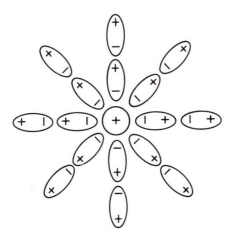

Figure 6.6. A positive charge in a liquid dielectric.

In the QED theory we assume that the same polarization effect occurs even in the vacuum where the virtual $e^- e^+$ pairs are polarized in an exactly analogous way to the molecules in the dielectric medium. Thus at large distances from the bare charge we see the Coulomb law behaviour, but at incredibly small distances we would expect to experience the full bare charge effects. In the normally attainable Coulomb region, the observed 'charge on the electron' is seen as the very large bare charge almost totally cancelled out by the vacuum polarization shielding effect. (This is related to the renormalization technique discussed in section 3.2.)

We can allow, approximately, for the QED vacuum polarization possibility by assuming that there is a modified fine structure constant that depends on momentum transfer in the following manner:

$$\alpha(q^2) \approx \frac{\alpha}{1 - (\alpha/3\pi)\ln(q^2/m_e^2)} \tag{6.3.1}$$

for $q^2 \gg m_e^2$ and where α is the normal fine structure constant. This formula sums the effects of simple vacuum polarization terms and shows that at large q^2 (small distances) the effective electromagnetic coupling is very slightly larger than the normal large-distance value of α. In general, such QED corrections make very small changes to predicted quantities but the remarkable fact is that very precise experiments have been carried out which agree with the theoretical calculations to a very high order.

We recall from chapter 3 that the coupling (fine structure) constant α is simply related to the charge on the electron by the relation

$$\alpha = e^2/4\pi\epsilon_0\hbar c = e^2/4\pi$$

in natural units.

Now, it is sometimes useful to think of the 'effective' charge on the electron, e_{eff}, in the same way, i.e.

$$\alpha(q^2) = e_{\text{eff}}^2/4\pi \tag{6.3.2}$$

and combining this definition with 6.3.1 we have

$$e_{\text{eff}}^2 = \frac{e^2}{1 - (e^2/12\pi^2)\ln(q^2/m_e^2)} \tag{6.3.3}$$

which tells us how the observed charge would depend upon q^2.

Let us turn our attention again to the QCD theory where the situation is even more interesting! There are now extra vacuum polarization terms as shown in Figure 6.7, arising from the allowed gluon–gluon couplings.

The 'quark-loop' diagram shown in Figure 6.7(b) acts in QCD in exactly the same way as the equivalent diagram in QED (Figure 6.7(a)). Thus the virtual $q\bar{q}$ pairs shield the colour charges on the scattering quarks just as the virtual e^+e^- pairs shielded the bare electron charge. However, the additional gluon diagrams (Figure 6.7(c and d)) act in the opposite direction, antishielding the colour charge, and it turns out that their effect greatly outweighs the shielding of the quark-loop diagram. There is thus a strong net antishielding effect which means that we have a dramatically different situation from the QED case.

We now have the colour force going to zero when the quarks are close together but becoming rapidly larger as the quarks are separated. It follows that

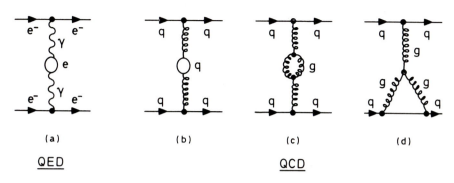

(a)   (b)   (c)   (d)

QED   QCD

Figure 6.7. Simple vacuum polarization diagrams.

at very large momentum transfers the quarks behave as if they are essentially free ('asymptotic freedom'). Quantitatively, we can express this effect in a similar way to the QED case by using a 'running coupling constant', α_s, which depends on q^2 as follows:

$$\alpha_s(q^2) = \frac{\alpha_s(q_0^2)}{1 + (\alpha_s(q_0^2)/4\pi)(11 - \tfrac{2}{3}N_f)\ln(q^2/q_0^2)} \tag{6.3.4}$$

where N_f is the number of quark flavours and q_0 is a 'cut-off' momentum transfer which is a scale parameter not explicitly given by the QCD theory. We see from this equation that provided $(11 - \tfrac{2}{3}N_f)$ is positive, that is, there are not more than 16 flavours of quarks (!), then as q^2 becomes very large ($q^2 \gg q_0^2$) the coupling α_s tends to zero, whereas if q^2 is very small ($q^2 \sim q_0^2$), then α_s tends to $\alpha_s(q_0^2)$. (If the number of flavours did indeed exceed 16 then the shielding effects of all those possible quark types would outweigh the antishielding of the 8 gluon varieties, and there would be a net shielding as in the QED case.)

By analogy with equation 6.3.2 in the QED case, we sometimes write

$$\alpha_s(q^2) = g_{\text{eff}}^2/4\pi \tag{6.3.5}$$

where g_{eff} is the effective colour charge coupling in the strong force.

Equation 6.3.4 is often written as

$$\alpha_s(q^2) = \frac{4\pi}{(11 - \tfrac{2}{3}N_f)\ln(q^2/\Lambda^2)} \tag{6.3.6}$$

where we have put

$$\ln \Lambda^2 = [\ln q_0^2 - 4\pi/(11 - \tfrac{2}{3}N_f)\alpha_s(q_0^2)]$$

In this form we see $\alpha_s(q^2)$ depends only on the one parameter, Λ, which is found experimentally to be of the order of a few hundreds of MeV/c^2 for regions of q^2

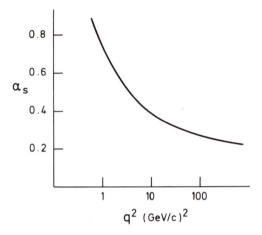

Figure 6.8. QCD running coupling constant.

which typically correspond to values of α_s ranging between $0{\cdot}2$ and $0{\cdot}8$ (see Figure 6.8).

An important aspect of this 'running coupling constant' idea is that it implies that in the nearly free regime α_s is sufficiently small that it makes sense to compute cross-sections using the same techniques as in QED. The higher order terms involving α_s^2, α_s^3, $\alpha_s^4 \ldots$ get progressively smaller so that 'perturbation' methods enable us to make reasonably good predictions for processes that occur at very high momentum transfers.

Another important point is that although equation 6.3.4 will not be valid for very small momentum transfers (large distances), it at least gives the hint that the force will be very large at distances of the order of 1 fm, giving us hope that we will eventually understand the origin of the bound states of quark systems and the reasons why only colourless particles seem to be physically observable. (Note that the weak coupling, g^2 (p. 35), also decreases with increasing q^2, but to a much smaller degree than we see in the strong force case. This decrease also arises from the fact that the intermediate vector bosons can couple to each other, like gluons, and this provides some antishielding effects.)

6.4. Confinement

It should be stressed that, so far, QCD has not shown that quarks and gluons are totally confined within the hadrons. We have the hint of an increasing colour force with distance but the region of extremely small q^2 is not accessible to QCD calculations as the value of α_s is too large to compute series with reducing terms in perturbation methods. It is sometimes called the 'non-perturbative region'. Certainly there is nothing in the current theory which forbids the existence of free quarks, although as we have mentioned before there is no compelling experimental evidence for their existence either. The attempt to prove that free coloured particles do not exist is a problem which is under very active study using a wide range of field-theoretical methods and models.

Even though we have no real dynamic theory for the bound states of quark systems there is a great deal of circumstantial evidence that our static model of the hadrons as q q q and q q̄ states is correct. The symmetries of the properties of the hadrons shown in chapter 4 only have been satisfactorily explained on this basis. It is useful to have a very crude picture in one's mind when thinking about these composite hadrons. It is obvious that an elastic string has the right general properties for representing the colour force. Two particles connected with such a piece of elastic are essentially free when they are much closer together than the unextended length of the string. When energy is put into the system to separate the particles the string starts to extend and the particles will feel an increasing force as they are driven further apart. If sufficient energy is put in, eventually the elastic breaks. In the model for quarks we would now have to assume that this energy is released and creates a new particle–antiparticle pair on the ends of the strings. This kind of model is illustrated in Figure 6.9 for the case of a meson in which we attempt to free a quark by pumping in energy. The process depicted in this figure ensures that only colourless composite quark states exist at all times. We simply end up with two mesons, the energy pumped in having been materialized into creation of the second particle.

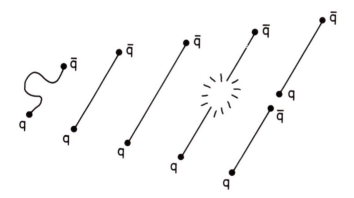

Figure 6.9. Attempting to separate a q q̄ pair in a meson.

The problem with this picture is the question of the origin of the 'elastic connection' between the quark and the antiquark. Let us go back to thinking about the gluons within the meson. As gluons can couple to gluons forming both a three-gluon vertex and a four-gluon vertex (see Figure 6.4), a meson will consist of a complicated mesh of gluons holding the quark and antiquark together (Figure 6.10(a)). In Figure 6.10 the 'springs' represent lines of colour flux carried

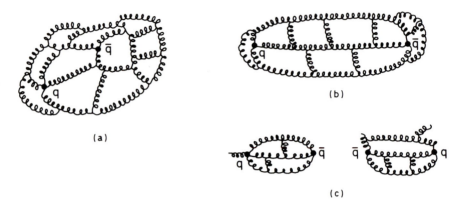

Figure 6.10. Colour flux lines in a meson.

by the gluons. In stretching the q and $\bar{q}$ apart we will have to assume that some-how the flux lines are squeezed into a tube-like shape as the separation takes place (Figure 6.10(*b*)). Eventually, when sufficient energy has been absorbed to drive the q and $\bar{q}$ beyond a certain distance apart, the tube will break leaving two colourless mesons, the energy having been turned into a second q$\bar{q}$ pair (Figure 6.10(*c*)).

If we take the tube radius as being independent of the separation distance between the q and $\bar{q}$, beyond some lower separation distance, then the potential increases linearly with distance. (We require this behaviour to fit in with our ideas about the energy levels of charmonium and beautyonium in which a linear potential is needed to give a good fit to the observed energy levels i.e. masses of the c$\bar{c}$ and b$\bar{b}$ confined states — see section 4.5.) However, there is a real problem with this picture. Why should the colour flux lines be squeezed into a tube as the separation increases? This certainly does not happen when you separate a pair of electric charges, where in fact the lines of electric force become more spread out as their distance apart is increased. Somehow then, there is a basic difference between the behaviour of electric and colour charge flux lines.

In trying to understand this last point, we are tempted to think of a bar magnet in which we try to isolate a 'monopole' (north or south pole) by pulling

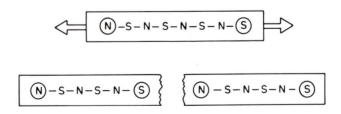

Figure 6.11. Trying to make a monopole!

the magnet apart into two pieces! It is clear that even if we are strong enough to effect the break we just end up with two bar magnets, each with its own pair of poles (see Figure 6.11).

To improve on this last model, consider putting a magnetic dipole into a cavity in a superconducting material. Now the situation is that the magnetic lines of force are indeed constrained so that they do not enter the superconducting medium. (This is known as the Meissner effect.) Thus, if the dipole is in a slot-shaped cavity, the lines of force will form themselves into a tube (Figure 6.12). This situation seems to be closely analogous to our model of the QCD colour flux lines. We are led to the idea that somehow the vacuum itself plays a part in constraining the colour lines of force to be within a tube. (Presumably this is the lowest energy, that is, most favoured solution.)

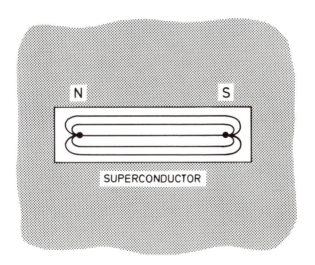

Figure 6.12. Magnetic dipoles inside cavities in a superconductor.

These pictures of 'confinement' mechanisms are not meant to be taken too seriously. Whilst it is clear that somehow or other the colour force indeed confines coloured objects within net colourless composite hadrons, we simply do not yet understand how it works. This is obviously a very unsatisfactory situation. It is as if we had the Coulomb law for the electromagnetic force, but could not calculate the wave-functions for the simplest bound states such as the hydrogen atom. For the same technical reason, we can only calculate a small fraction of all the possible classes of hadronic collisions that can occur in scattering processes. Only the high transverse momentum scatters which are dominated by scatters between individual quarks and gluons (the so-called 'hard scatters') are calculable. It seems as though we are still some way from any sort of complete understanding of the strong interaction.

6.5. *'Valence' and 'sea' quarks*

The quarks and antiquarks that we have been discussing in the context of
qqq and $q\bar{q}$ hadronic states give the hadrons their characteristic properties and
are called the 'valence' quarks. This is by direct analogy with atomic structures in
which the valence electrons in the outer shells give atoms their chemical
properties.

However, the $q\bar{q}$ pairs that come out of the vacuum as in Figure 6.7(*b*), that
we were discussing in the context of vacuum polarization in section 6.3, will also
play a part in the properties of a hadron, such as a proton. Although the proton's
main properties are determined by the u, u and d valence quarks, occasionally one
of the gluons within the proton can briefly transform to a $q\bar{q}$ pair (see Figure 6.13)
and sometimes a high-energy probe will detect their existence. Such additional
particles are referred to as 'sea' quarks as the valence quarks are, as it were,
swimming in a sea of virtual $q\bar{q}$ pairs. There is therefore a finite probability that a
high energy electron probing the proton at very small distances will sometimes
interact with an antiquark even though we normally think of the baryon as consisting of three quarks. The results of experiments that probe the proton actually
suggest that a few percent of the internal momentum of the proton originates
from the sea quarks and antiquarks.

Our picture of a proton has now become considerably more complicated. We
have our u, u and d valence quarks swimming in a sea of $q\bar{q}$ pairs (mostly the
lighter ones, but sometimes $s\bar{s}$ or even $c\bar{c}$ pairs) all interconnected by a complex
mesh of gluons (see Figure 6.13).

6.6. *Tests for QCD — 'hard' scattering*

We have already discussed some of the evidence for the idea that we need a
new quantum number corresponding to the colour charge. We have also said that

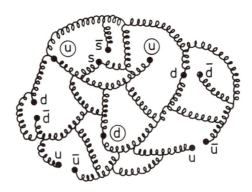

Figure 6.13. The proton with its valence and sea quarks as well as the gluons.

the QCD theory has been quite successfully applied to the calculation of cross-sections for the processes which are dominated by interactions between the hadronic constituents (quarks and gluons). These calculated hard scattering cross-sections are typically in agreement with measured values to within a few tens of percent. As the higher order corrections are relatively difficult to compute this measure of agreement is regarded as a success for the theory.

There are, however, two crucial tests of the theory. First, the existence of gluons must be established and, secondly, we need to see evidence for the running nature of the colour coupling, that is, the q^2-dependence of α_s).

Some of the scattering experiments carried out in the late 1960s suggested the idea that there are neutral constituents in nucleons, in addition to quarks, and that up to half the momentum of the nucleon is carried by these neutral objects. However, the most direct evidence for the existence of gluons has come from recent experiments at the PETRA collider at DESY. Hadrons were produced in $e^+ - e^-$ collisions by colliding electron and positron beams at very high energies. The hadrons were found often to be produced in 'jets' of particles. When such jets are observed they are usually seen in pairs. Their interpretation is that they mainly come from direct production of quark–antiquark pairs which 'dress themselves up' with other quarks and antiquarks to form showers of observable hadrons which come out clustered together in space (see Figure 6.14(*a*)). This 'dressing up' process (called fragmentation) occurs because the free coloured quarks cannot be released from the collision and some of the energy of the collision must be used to pick up $q\bar{q}$ pairs from the vacuum so that colourless hadron states can be emitted into the laboratory. The observed angular distribution of the axes of the jets is just what would be expected from two spin 1/2 particles, nicely confirming the idea that they originate from the produced quark and antiquark. The most interesting events, however, appear to have three such jets. It is assumed that we are then observing a quark–antiquark pair production, but in addition one of the quarks has emitted a gluon which fragments into a third observable jet of hadrons (see Figures 6.14(*b*) and 6.15). (This gluon radiation

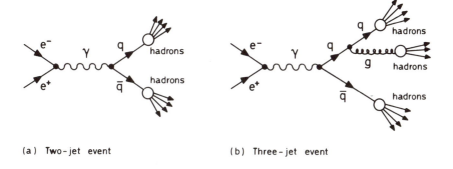

(a) Two-jet event   (b) Three-jet event

Figure 6.14. Hadronic production in $e^+ - e^-$ collisions.

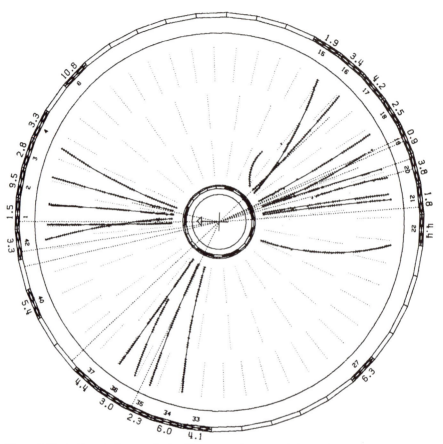

Figure 6.15. A 'three-jet' event obtained in an experiment carried out with the PETRA collider at DESY. The origin of the jets is thought to be the production of a quark, antiquark and gluon in the e^+–e^- collision.

process is analogous to the emission of a photon from an accelerated electron.) Detailed analysis of these three-jet events is at least consistent with this picture.

Evidence for the q^2-dependence of the strong coupling constant is seen in the q^2-dependence of the distribution of momenta of the quarks within the hadrons. Such a distribution is called a structure function and can be explored by probing the hadron with a high-energy particle; typically electrons or neutrinos are often used in such studies. The observed shape of the distribution is found to depend quite strongly on the momentum transfer involved in the collision of the probe with the hadron (see Figure 6.16).

One striking result of such structure function studies is that if we sum the momenta carried by the quarks in a hadron determined from scattering experiments, we can account for only about half of the momentum of the hadron. As

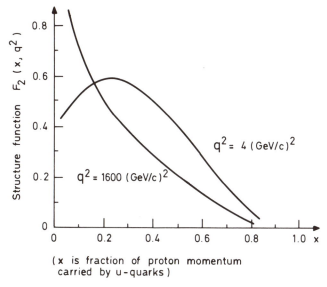

Figure 6.16. Structure function of the nucleon at various q^2.

mentioned before, it looks as though the other half is carried by the gluons.

Overall, it seems that the experimental evidence for the specific approach of QCD is very striking, but there are still some worrying problems with the theory, mostly concerned with bound states and low-q^2 scattering.

6.7. 'Soft' scattering

Most of the quantitative tests of QCD that have been discussed in the last section are based on data taken in high-transverse-momentum collisions where the quarks are nearly free and perturbative calculations of various cross-sections can be carried out. However, by far the bulk of collisions that occur when a beam of high-energy protons hits a liquid hydrogen target happen in such a way that a small transfer of momentum takes place. In these low-q^2 ('soft') scatters, it is the collective behaviour of the whole proton which determines the details of the interaction. It is presumably the residual colour force which determines the cross-sections. In just the same kind of way, it is the Van der Waal force which determines the behaviour of molecular collisions, and this is a residual coulomb basic force from all the individual molecular constituents.

In these soft collisions, the hadrons may hold together, in which case we have an elastic scatter, or they may break up and produce new particles if there is sufficient energy for the process to occur. Either way, it is the overall collective properties of the hadrons which dominate. Many such collisions can be understood in terms of very simple geometrical models in which the wave associated with the incoming particle is scattered by a nearly opaque disc-shaped object which represents the target proton. We do indeed observe a narrow peak in the cross-section in the forward direction followed by other maxima and minima, which is very reminiscent of optical diffraction patterns from light incident on a black disc.

The original ideas of the strong force (the Yukawa theory) were based on another very simple model in which the exchanged particle was a pion. This single-pion-exchange model (see Figure 6.17) gives about the right range for the strong force and qualitatively explains many of the features of the low-q^2 hadronic scattering. However, to get a more complete picture it was necessary to consider the exchange of many other states (such as the ϱ, the ω and the ϕ).

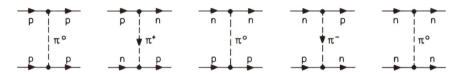

Figure 6.17. Single pion exchange model for p–p, n–p and n–n scattering.

For many years, a much more complicated development of these ideas, known as the Regge exchange mechanism, was paramount. We will not discuss such theories as they probably do not help us to understand the basic mechanisms, even though they may give quite useful and economic descriptions of the data.

CHAPTER 7
speculations and problems

7.1. Grand unified theories (GUTs)

One of the main aims of fundamental physics, perhaps the only aim, is to describe as many physical phenomena as possible in the most economic manner. From this point of view we would like to minimize the number of laws of force and the number of arbitrary parameters that appear in the theories.

Maxwell took the whole range of electrical and magnetic data accumulated from the experiments of Faraday, Ampère and many others and achieved the unification of description of these phenomena which he was able to express through his famous set of four equations. His theory also had great predictive power. In particular, it predicted the properties of the whole range of electromagnetic radiation and unified our understanding of radio waves, microwaves, the infrared, visible and ultraviolet radiations, X-rays and the very high frequency λ-rays. These Maxwell equations thus encompass an enormous range of experimental data with great economy.

With the addition of special relativity and the ideas of quantum field theory, we now have the most complete theory of the electromagnetic processes embodied in quantum electrodynamics (QED). Within this theory we can also incorporate another whole range of phenomena such as atomic spectra, the Compton effect, the photoelectric effect, black-body radiation and the most basic electromagnetic scattering processes discussed in earlier chapters.

In chapter 5 we saw how the next step in the saga of unification included the weak interaction phenomena into a common framework with the electromagnetic effects, in the new electroweak theory. Its very specific predictions of neutral currents and the existence and masses of the intermediate vector bosons have been brilliantly confirmed experimentally.

The next step seems obvious. Why should we not also try to include the strong colour force theory (QCD) into a common scheme with the other two forces already unified? (Ideally we would like to bring in gravity as well.) It seems that there are some general similarities in all the models we have been considering. In every case we have described the interactions as occurring through the exchange of field bosons and the particles which 'feel' the forces as carrying some kind of 'charge'. (These charges are mass, weak charge, electric charge and colour charge for the gravitational, weak, electromagnetic and strong forces, respectively.) The different strengths of the forces are determined by these charges.

There is a very much deeper, but unfortunately rather technical, similarity in the basic mathematical structure of the various theories. In all cases we assume that we are dealing with 'gauge' forces which is a concept that has already been discussed in chapter 5. It is the gauge symmetries which link the strengths of an interaction to a charge. We have seen that the weak forces involve a broken symmetry, the weak charge is not conserved and the masses of some exchanged bosons are non-zero. On the other hand, the colour symmetry seems to be exact, but nevertheless the gauge theory is of the same class (Yang–Mills). A 'grand unified theory' ('GUT') incorporating electroweak and strong forces will be presumably contained within a broader symmetry in the same class of gauge theory and must include both the colour-symmetric and broken flavour-symmetric aspects.

Of course, the real question is how can we possibly include into one theory phenomena that occur with such widely different probabilities (or strengths)? These coupling strengths vary only very slowly with energy but it seems possible that they could be truly unified at some enormous energy. The electromagnetic and weak unification occurs at an energy of around 100 GeV and making the necessary huge extrapolation (see Figure 7.1) it looks as though the unification with the colour force could happen at about 10^{14} or 10^{15} GeV.† On any standards this is speculative!

Ideally, we would like to have just one coupling constant at some very high energy ($\sim 10^{15}$ GeV) and what we actually observe at 'normal' energies is then determined by three aspects of what is fundamentally a single basic interaction. At energies of around 100 GeV the weak interaction strength has come up to meet the electromagnetic strength, but we have to go to 13 orders of magnitude higher energies to bring the colour force down to the common strength of the other two forces. Should it be possible ever to reach such energies we would not then expect there to be any distinction between hadrons and leptons, all the fundamental constituents seeing the same common force. Although it seems to be quite out of the question to develop any artificial means of reaching such huge energies, processes at these energies must certainly have been very important in the early stages of development of the Universe and an understanding of the unification of forces is a central requirement for cosmological theories.

At the grand unified energy, it is postulated that there is a new symmetry between the quarks and leptons and quark–lepton transitions can occur which will imply the breaking of the conservation of baryon and lepton numbers. Such transitions could occur through the virtual exchange, or the production and subsequent decay, of new very high mass ($\sim 10^{15}$ GeV/c^2) bosons with spin-1 called

---

† Note that 10^{15} GeV $\doteq 1 \cdot 6 \times 10^5$ J, which is about equal to the kinetic energy of the 10^{30} particles in a car moving at 50 mph. Our difficulty is that this is the energy needed by a single subnuclear particle to reach the unification energy.

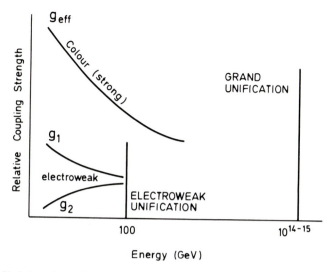

Figure 7.1. Variation of coupling strengths with energy.

the 'X' and 'Y' or sometimes the 'lepto-quarks'. Thus at sufficiently high energies the processes depicted in Figure 7.2 can break the conservation rule for baryon number, B. For example, the X particle could sometimes decay into an anti-lepton–antiquark and sometimes into a pair of quarks and the relative probabil-

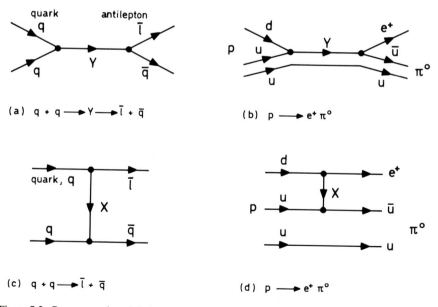

Figure 7.2. Baryon number violation.

ities for these two decays will determine the extent of the violation of the baryon number conservation rule.

The baryon number violation process would also break the charge conjugation (C) symmetry and would lead to a particle–antiparticle asymmetry. This is a particular point of interest in seeing the relevance of grand unified theories to understanding the early development of the Universe. Following the great energy explosion (the 'Big Bang'), the immediate evolution of the Universe must depend crucially on the behaviour of particles at enormously high energies. The balance of the initial creation of particle–antiparticle pairs and the subsequent annihilations must have been upset at an early stage by such baryon number violating possibilities. In particular, we have to explain the fact that some process allows the existence of 10^9 photons to each other particle that enables the rest of the development of the Universe to be at least partially understood. The grand unified theory can be tuned to give just the required degree of violation with the hypothesis of the lepto-quark particle to allow just such a ratio of photons to particles.

Particle physicists obviously are interested in the possibility of observing effects at attainable energies which depend upon the existence of such grand unified theories. One possible line of experimental enquiry is to look for the decay of the proton. If we consider the interaction shown in the Feynman diagram of Figure 7.2(*b*) we see that there is a predicted process which allows three quarks (as in a proton) to go to a q q̄ pair (which could be a neutral pion or kaon) and an anti-lepton (such as a positron or positive muon). The grand unified theories thus predict that the nucleons will decay through such processes as

$$p \rightarrow e^+ + \pi^0 \tag{7.1.1}$$
$$\text{or } p \rightarrow e^+ + K^0 \tag{7.1.2}$$
$$\text{or } p \rightarrow \bar{\nu} + K^+ \tag{7.1.3}$$
$$\text{or } n \rightarrow e^+ + \pi^- \tag{7.1.4}$$

and many other possible channels.

The grand unified theories thus make the startling prediction that the proton may not be stable but may decay, for example, with the emission of a positron. It is obviously not a common process or otherwise, since we are mostly made of protons ourselves, we would not last very long! In fact, it has been estimated that as we seem not to be decaying to any measurable degree, there must be a lower limit on the lifetime of the proton at a level of about 10^{16} years.

Bearing in mind that the lifetime of the Universe to date is about 10^{10} years, we are clearly dealing with immensely long lifetimes and correspondingly minute decay probabilities.

In a simple GUT model due to Georgi and Glashow, the prediction for the rate of decay of the proton, Γ_p, is as follows:

$$\Gamma_p \sim \alpha_G^2 \, m_p^5 / m_X^4 \tag{7.1.5}$$

where α_G is the grand unified coupling constant and m_X is the mass of the X-particle. If we put in a mass of $10^{15}\,\text{GeV}/c^2$, and if α_G is of order 1, then the lifetime is of the order of 10^{30} years. More detailed calculations for the process 7.1.1, for example, give an estimate of about $2 \cdot 3 \times 10^{31}$ years. How can one possibly measure such a lifetime when the times involved are more than 10^{20} times the age of the Universe? The answer, of course, is to study a very large mass of material and make use of the fact that Avogadro's number is also very large! For example, if we look at 1000 tons of water for a few years we should see perhaps five examples of a proton decaying into a pion and a positron.

A number of experiments around the world are looking for evidence for the decay of protons. For example, the experiment of the IMB collaboration (Irvine–Michigan–Brookhaven) has looked at 2300 ton.years of material and is setting lifetime limits of the order of 10^{32} years for the decay process 7.1.1. There is as yet no convincing evidence for the direct observation of proton decay, but the data from a number of experiments are not inconsistent with a signal of the order of a few times 10^{32} years. The experimental limitations arise from the backgrounds resulting from cosmic-ray neutrino interactions and it will be very difficult ever to measure lifetimes longer than about 10^{33} years. It does seem, however, that the existing data rule out the very simplest GUTs models which predict rather shorter lifetimes, but there are many uncertainties in both the theories and the experiments. It is relatively early days for such conclusions to be made, and proton decay measurements are surely going to be among the most crucial experiments in the near future.

Apart from the direct observation of proton decays, it is very difficult to see how we can obtain evidence for grand unification. The present accelerators are reaching energies in the hundreds of GeV range and within the foreseeable future we may reach nearly 10^5 GeV. However, this is a very long way off 10^{15} GeV where the full effects of unification can be expected.

On the other hand, grand unification plays such an important role in the understanding of the early Universe that we may obtain more indirect insight into GUTs theories from astrophysical and cosmological studies. This is one of a number of areas in which particle physics is playing an interesting part in the understanding of other fields of physics.

The grand unified theories do, however, have some real problems. They certainly unify the parts of the standard model into a single framework. Unfortunately, there is very little reduction in the large number of arbitrary parameters involved in these theories compared with those of the standard model. It is also a little disturbing that there are two such very different energy scales involved, one around the mass of the W and Z particles ($100\,\text{GeV}/c^2$) and the unification energy around 10^{15} GeV. Moreover, the GUTs models do not seem to help in bringing gravity into the picture. We expect quantum gravitational effects to be extremely important at energies that correspond to the 'Planck mass' at about $10^{19}\,\text{GeV}/c^2$. (The Planck mass is equal to $(\hbar c/G)^{1/2}$ where G is the Newton universal constant of gravitation, and at such energies quantum effects will be significant.) One attempt to go further involves a new idea referred to as supersymmetry, which is a symmetry connecting fermions and bosons.

7.2. Supersymmetry (SUSY)

Many of the symmetries we have so far considered have had the effect of grouping particles into multiplets. For example, the isospin symmetry (see section 2.5) groups together particles with the same spin and nearly the same mass, but with differing electric charges. Isospin transformations change members of a particular isospin multiplet into other members of the same multiplet. The underlying physical origin of this symmetry is simply that the forces arising from the colour charges are independent of the electric charges. When we introduced the strange quark (s) into the model, we found larger multiplets in which the members have the same spin state and very roughly similar masses (for example, the spin 3/2 baryon decuplet shown in Figure 4.2(b)). It is almost as if the members of a multiplet are simply different facets of the 'same' particle whose different attributes are presented to us in the different members of the set. (Of course, the fact that the masses are not exactly the same indicates that the symmetry is not exact, but is broken.)

With the same approach, some theoretical physicists are attracted to the idea that much larger multiplets can be formed if we assume there is a symmetry between fermions and bosons. Supersymmetry will thus transform bosons into fermions and fermions into bosons, so grouping together particles and the mediators of the forces into large groupings called 'supermultiplets'. In SUSY all particles have a supersymmetric partner. The current nomenclature is that the bosonic partner of a fermion adds an 's' in front of its name, while the fermionic partner of a boson adds '-ino' to its root. Thus the spin 0 partner of the $s = 1/2$ quark is called a 'squark' and the spin 1/2 partner of the spin 1 gluon is called a 'gluino'. Other examples of supersymmetric partners are given in Table 7.1.

Table 7.1. Supersymmetric partners.

Particle	Spin	Supersymmetric partner	Spin
quark	$\frac{1}{2}$	s quark	0
lepton	$\frac{1}{2}$	s lepton	0
W	1	Wino	1/2
Z	1	Zino	1/2
gluon	1	gluino	1/2
photon	1	photino	1/2
graviton	2	gravitino	3/2

If this supersymmetry were to be exact then the masses of the partners would be identical to the normal particles. However, it is already clear experimentally that if these new particles exist they must have rather large masses (typically greater than $20\,\text{GeV}/c^2$ in most cases). So far, there is not a shred of evidence for the existence of supersymmetric particles!

However, supersymmetry is an attractive idea theoretically as not only does it have a certain aesthetic appeal, it also enables gravity to enter the model. SUSY is obviously a strongly broken symmetry but the question is, what kind of mechanism is responsible for the effect? There could be a spontaneous symmetry breaking mechanism analogous to that which occurs in the electroweak theory. Alternatively, the breaking of supersymmetry can come about through its inter-action with gravity. Such a 'supergravity' (SUGY) theory predicts a massless 'gravitino' with spin 3/2 as the supersymmetric partner of the spin 2 massless graviton which is postulated to be the carrier of the gravitational force. It is really the imposition of local supersymmetry (see section 5.2) which requires these two new fields corresponding to the spin-2 graviton and the spin-3/2 gravitino (that is, local SUSY is SUGY!). All of this assumes that gravity has a quantum nature, which is by no means obvious. The gravitational force has been brought in through the introduction of a gauge invariance into supersymmetry.

Again one must ask what experimental tests can be made which are specific to supersymmetry as opposed to, say, GUTs. The obvious point is that there should be a wealth of new particles, probably with large masses. In fact, there should be one supersymmetric partner for every known fundamental constituent. In addition, there should be new observable colourless composite particles con-taining one or more of the new constituents. For example, it has been postulated that 'R-hadrons' should exist which would be normal baryons or mesons but which would contain an additional gluino. An R-pion would thus contain a $q\bar{q}$ pair and a single gluino and it is estimated that such a particle would be fairly long-lived (a lifetime of perhaps 10^{-15} to $10^{-12}\,\text{s}$) decaying into a photino and hadrons. Various searches for such objects are now under way, as are searches for many of the other predicted supersymmetric states. Proton decays can also be accommodated within this model, but the dominant decay mode would be into a kaon and an antineutrino and at present we have no firm evidence for such decays either.

7.3. A quark and lepton sub-structure?

The standard model that we have described in the earlier chapters gives a very good description of the presently known behaviour of fundamental particles and a good model for the nature of the forces between them. If we look again at Table 4.9 we can remind ourselves of the basic constituents required by the standard model. There are 24 fermions (6 quarks, 6 antiquarks, 6 leptons and 6 antileptons) as well as the 12 field bosons. Why should we need 24 fermions? After all, we know that ordinary material can be 'understood' in terms of just three fermions (the u and d quarks and the electron). Notice that the most common fermions are all members of the first generation. What could be the function of the second and third generations, and are there perhaps even more generations whose members have not yet been discovered? An earlier and much limited version of this worry used to be known as the muon problem, as nobody could find a role for the muon. A closely related problem is the wide range of masses of the various constituents which contributes to the large number of arbitrary parameters which have to be artificially inserted into the standard model.

If we try to learn from past experience, we notice that physicists have been in this situation several times before. For example, at one time atoms were thought of as being 'elementary' particles and the regularities of their properties seen in the periodic table eventually gave rise to an understanding of the symmetries involved and a dynamic model of atomic structure emerged. This model, together with rather similar developments in the understanding of the structure of nuclei, reduced the number of material building blocks to just three; namely, protons, neutrons and electrons. Moreover, the dynamic model predicted that there should be excited states of atoms that are allowed by the quantum mechanical rules given that enough energy is injected into the system. A very analogous series of steps occurred in understanding the nature of hadrons. In the decade starting from the mid-1950s, large numbers of complicated hadronic states were observed and their properties were classified. Again the symmetries of these properties led to the simple quark model, greatly reducing the number of particles that we regarded as 'elementary'. As in the atomic example, it was quickly realized that many of the observed hadrons were unstable and were just excited states of lower mass particles in the sense that all their properties were identical to the ground-state particle except for the mass (energy) of the state and its stability.

The obvious implication in the present situation is that we should try to simplify the quark and lepton standard model by postulating an inner structure of some more 'elementary' building blocks which can reflect the observed symmetry of the first three generations of quarks and leptons that we see in Table 4.9. The most natural assumption would appear to be that the second and third generation particles are just excited states of the first-generation corresponding ground states. Each high-generation member seems to have identical behaviour in the way it interacts compared to the corresponding first-generation constituent, except in so far as its higher mass affects the kinematics of the situation. In this picture the s and b quarks, for example, are excited versions of the d quark and the μ and τ leptons are excited states of the electron. Other theorists hold the view that all 24 fermions are on the same ground-state footing and they look for models that have a rather more complicated sub-structure. In these models, you still expect excited states of the electron, for example, but these will be additional to the presently known heavy-mass leptons.

The notion of an inner structure of quarks and leptons has an immediate appeal as an obvious simplification. If we make the bold assumption that the same inner constituents contribute to both quarks and leptons then an under-standing of the forces between them should also give us another approach to grand unification as the sub-particles, sometimes called 'preons' or 'prequarks', would cross the quark–lepton boundary and could unify fermions and bosons in their various interactions.

Despite the strong aesthetic appeal of such a theoretical direction, there are some obvious and very real difficulties. Many theorists have postulated such preon models, but all of them flounder on a simple concept. Experimentally, we know that if such preons exist they must at least be confined within a region which is less than 10^{-16} cm. We know from scattering experiments that quarks occupy a region in the hadron which is of the order of $1/1000$ of the diameter of the proton or less. Indirectly, we also know that the electron also occupies a similar sized region, or less. In QED calculations of quantities such as the magnetic moment of the electron, the electron is assumed to be 'point-like' and very high-order calculations, together with correspondingly high precision experimental measurements, have set limits at the 10^{-16} cm level. Thus we are fairly clear that if preons exist and form composite quarks and leptons they must have 'diameters' of the order of 10^{-16} cm or, probably, much less. A simple application of the uncertainty principle immediately tells you that distances of this order correspond to energies in the hundreds of GeV range or more. Thus, we have to accept that the energy of the preons is enormously larger than the mass of the composite objects that they form (the quarks and leptons).

We are therefore looking at a very different situation to, say, the hydrogen atom consisting of a proton and an electron. The mass of the whole atom is only just less than the sum of the masses of its constituents, the difference being the binding energy. In nuclei we have a similar situation where the so-called 'mass defects' are typically about one per cent of the masses of the nuclei. The question that arises, then, is how can the preons be bound in such a way as almost to cancel out their enormous intrinsic energies? This is an unsolved problem which many theorists are attacking. It is perhaps particularly difficult to think of neutrinos as being composites with their zero (?) mass.

Nevertheless, although we do not yet have a satisfactory dynamic model for inner constituents of quarks and leptons, there have been many quite successful static models in which the basic regularities in the properties of quarks and leptons are obtained from a set of preons.

As an example, let us consider a particularly simple approach used by Haim Harari (and others) in the 'rishon' model. (Rishon is the Hebrew word for primary.) He assumes that there are just two prequarks with electric charges $1/3$ and 0 (T and V, respectively) and two antiprequarks with electric charges $-1/3$ and 0 ($\overline{T}$ and $\overline{V}$, respectively). In this model the Ts and $\overline{V}$s can have any of the three colour charges (R, G and B) and the Vs and $\overline{T}$s take any of the three anti-colours ($\overline{R}$, $\overline{G}$ or $\overline{B}$, see Table 7.2). The first generation quarks and leptons are then simply constructed by taking any three rishons or any three antirishons, without mixing them, in the same composite particle (see Table 7.3).

Several nice features of this model emerge. First, we have only two basic particles together with their antiparticles. We see that all integrally charged

Table 7.2. Charges on the rishons.

Prequark	Electric charge	Colour charge
Rishon, T	$+1/3$	R, G or B
Rishon, V	0	$\bar{R}$, $\bar{G}$ or $\bar{B}$
Antirishon, $\bar{T}$	$-1/3$	$\bar{R}$, $\bar{G}$ or $\bar{B}$
Antirishon, $\bar{V}$	0	R, G or B

Table 7.3. Rishon model of first generation standard model constituents.

First generation composite particle	Prequarks	Rishon content Electric charge on prequarks	net	Rishon content Colour charge on prequarks	net
e^-	$\bar{T}$, $\bar{T}$, $\bar{T}$	$-1/3$, $-1/3$, $-1/3$	-1	$\bar{R}$, $\bar{G}$, $\bar{B}$	white
$\bar{u}$	$\bar{T}$, $\bar{T}$, $\bar{V}$	$-1/3$, $-1/3$, 0	$-2/3$	$\bar{B}$, $\bar{R}$, R	$\bar{B}$
				or $\bar{B}$, $\bar{G}$, G	
				$\bar{G}$, $\bar{B}$, B	$\bar{G}$
				or $\bar{G}$, $\bar{R}$, R	
				$\bar{R}$, $\bar{G}$, G	$\bar{R}$
				or $\bar{R}$, $\bar{B}$, B	
d	$\bar{T}$, $\bar{V}$, $\bar{V}$	$-1/3$, 0, 0	$-1/3$	(as for u)	B, G or R
ν_e	$\bar{V}$, $\bar{V}$, $\bar{V}$	0, 0, 0	0	R, G, B	white
e^+	T, T, T	$+1/3$, $+1/3$, $+1/3$	$+1$	R, G, B	white
u	T, T, V	$+1/3$, $+1/3$, 0	$+2/3$	B, G, $\bar{G}$	B
				or B, R, $\bar{R}$	
				G, R, $\bar{R}$	G
				or G, B, $\bar{B}$	
				R, G, $\bar{G}$	R
				or R, B, $\bar{B}$	
$\bar{d}$	T, V, V	$+1/3$, 0, 0	$+1/3$	(as for u)	B, G or R
$\bar{\nu}_e$	V, V, V	0, 0, 0	0	$\bar{B}$, $\bar{G}$, $\bar{R}$	white

particles are colourless and all fractionally charged particles are coloured. In this framework, it is also clear why the charge on the proton is equal and opposite to that on the electron, so ensuring that the hydrogen atom, for example, is exactly electrically neutral. The proton consists of two u quarks and one d quark. From Table 7.3 we see that the total number of preons is:

$$4\,T + 2\,V + 1\,\bar{T} + 2\,\bar{V}$$
with charges $4/3 + 0 - 1/3 + 0 = +1$

The electron is formed from three $\bar{T}$'s with electric charge $3 \times (-1/3) = -1$ and so the atom is exactly neutral. Hence, the basic charge and colour charge properties are properly satisfied.

In this particular model, the higher generations have to be regarded as excited states of the first-generation constituents, so that one might expect rare decay processes like

$$\mu^{\pm} \rightarrow e^{\pm} + \gamma$$

to be observed, whereas in fact only upper limits have been obtained. No proper dynamic model for such higher energy states exists, mainly because of the problem of the binding energy discussed above. The hypothesis has been made that the preons are held together by a 'hypercolour' force, by analogy with the normal strong colour force, which is mediated by 'hypergluons' (or perhaps superglue!) acting between the preons. Presumably there is a hyper-confinement process which means that only hypercolourless states (that is, quarks and leptons) can be observed but that these composite particles can carry normal colour.

Other static preon models, such as that suggested by Pati and Salam, get over the problem of the higher generations by assuming that some preons carry a new quantum number of generation, some carry electric charge and some carry colour charge. The snag with such an approach is that you end up with a rather uncomfortably large number of preons. In the Pati–Salam model there are nine basic states, and new excited states of the composite particles would be expected, such as excited electrons.

In conclusion then, there are successful static preon schemes but a full dynamic model of an inner structure for quarks and leptons seems some way off. It should also be noted that the observation of proton decay can perfectly well be accommodated within these models, as obviously the preon schemes mix leptonic and hadron fermions, allowing for the breaking of baryon number conservation.

7.4. Unanswered questions and conclusions

Let us now try to see where we stand in the attempt to understand the pattern of fundamental particles and the forces between them. First, we can say that the standard model has had a great success in providing an overall scheme for the observed particles and their various possible interactions. We will consider each main aspect of the model; namely, QED, electroweak theory and QCD.

QED has had a long and highly respectable history and continues to be one of the mainstays of theoretical particle physics. It covers the one area that we are relatively sure is well understood. Theory and experiment go hand-in-hand, providing more and more precise predictions which are tested with better and better experimental measurements. Crucial, mainly low-energy, tests have been carried out and agreement to ten significant figures has been achieved. There is, however, always room for even better measurements of such quantities as the magnetic moments of the electron and the muon which will give even tighter limits on such interesting questions as whether or not the leptons have a point-like nature.

The electroweak theory has also been remarkably successful in its very specific predictions of neutral currents and the masses of the W and Z particles. There is, for the first time, a convincing connection between the weak and electromagnetic phenomena. There are, nevertheless, some problems associated with this theory and as a consequence it does not yet have quite the same firm footing as QED. The main stumbling block is the exact nature of the mechanism for breaking the gauge symmetry. It is clearly broken, as we have three massive intermediate vector bosons (W^+, W^- and Z^0) and one massless photon. In QED there is an exact gauge symmetry which is reflected in the zero mass photon and within QCD there are similarly massless gluons associated with the exact colour symmetry. This problem forces the Higgs mechanism into the electroweak theory and the Higgs bosons break the symmetry spontaneously, giving mass to the W and Z particles (and the fermions) but there should be at least one Higgs not 'used up' in this process as the photon is massless. So far, we have no hint of any experimental evidence for such a Higgs boson. (Of course, it may be simply that the mass of the Higgs particle has so far precluded us from producing it with the available accelerator energies.)

Turning now to the strong force, QCD is clearly the best theory available to us, but as we have indicated before it is difficult to use this theory to calculate many of the strong interaction processes. As the coupling is large at small momentum transfers, the large cross-section soft scattering processes are a particular problem. Even with the hard scatters, where the quarks are essentially free, there are technical difficulties in making very precise high-order calculations and certainly nothing like the precision of QED seems to be attainable. Furthermore, we cannot really convincingly cope with colourless bound states of quarks until we have a full understanding of the confinement mechanism. For the moment, it is not clear whether confinement is exact or only approximate. The possibility of observing free quarks is certainly not yet ruled out.

Some preliminary evidence has very recently been obtained for the existence of the 'top' quark with a mass of between 30 and $50 \, \text{GeV}/c^2$. This has come from a few events in the p–p̄ collider at CERN in experiment UA1. It is obviously important to confirm this result and obtain an accurate mass for this final member of the third generation of fundamental constituents. We can expect this in the near future with the expected improvements in the p–p̄ collider system at CERN.

Probably the biggest problem arises from the large number of quarks which are seen in the standard model. What is their function and where do their masses come from? In fact, in the standard model as a whole, the number of particles with a wide range of masses is an embarrassment. For many years physicists worried about the role of the muon which seemed to have no purpose. Now we see the whole second and third generations of particles as being essentially superfluous!

Another interesting question arises from the connection between the electric charges of the constituents. In each generation we see all possible charges in steps of 1/3 from -1 to $+1$. For example, in the first generation we have:

$$-1 \quad -2/3 \quad -1/3 \quad 0 \quad +1/3 \quad +2/3 \quad +1$$

$$e^- \quad \bar{u} \quad d \quad \frac{\nu_e}{\bar{\nu_e}} \quad \bar{d} \quad u \quad e^+$$

Why do we have just those values and no others? For example, why do we not have $-4/3$ or $+11/3$? A related, and older, problem is the question of why the proton seems to have exactly the same magnitude of electric charge as that carried by the electron.

The answer to some of these questions could come from an underlying sub-structure to the quarks and leptons. As we have seen in section 7.3, there are as yet no respectable dynamic models of this type, although the general idea seems very attractive. Quarks and leptons may not be the only fundamental particles that could have sub-structures. Some physicists feel that the very fact that the W's and Z have such large masses may already be an indication that they are also composites, although again there is no theory for the binding in these cases. It is possible that direct evidence for new excited states of apparently fundamental particles such as the electron could be found. For example, we might find such evidence in the decays of W and Z^0 particles when they are produced abundantly by the next round of accelerators. The Z^0 could decay into an excited positron and an electron, for example, and the excited state might then drop down to the normal positron with the emission of a photon. However, all of this is pure conjecture and we must await new experimental evidence.

In considering grand unified theories, we might have hoped that not only would there be a better understanding of the nature of the forces, but also that there would be a reduction in the number of arbitrary parameters that have to be provided. It is disappointing that little progress seems to be made in this latter direction compared to the standard model situation. The other worry with at least the simpler versions of GUTs is that there are clear predictions of the proton lifetime which have probably already been surpassed by experimental limits. For the moment, proton decay is about the only testable prediction of GUTs at available energies and until a clear signal for such decays is obtained, little progress can be made. (Notice that such decays are also compatible with preon models and supersymmetry as well as grand unification.)

The other step beyond the standard model that we have discussed is the introduction of fermion–boson supersymmetry (SUSY). Here again, despite the elegance of the theory, at present there is just no evidence to support the basic idea. Many experimental searches for such evidence are under way, but the present indication is that if supersymmetric partners exist, then they probably have masses that are too high to be produced copiously at the present laboratories. The question of the introduction of quantum gravity into a unified theory is one of the major frontier areas of theoretical particle physics and the super-gravity models are seen to be an attractive option within the supersymmetric theories.

Table 7.4. Some of the world's approved and proposed new colliders.

Name		Particles accelerated	Energy	Planned start-up date
Western Europe				
CERN	Large Electron Positron Collider, LEP I (see Figure 7.3)	e^+ e^-	50 GeV on 50 GeV	1989
CERN	LEP II	e^+ e^-	100 GeV on 100 GeV	1992
CERN	Large Hadron Collider LHC	p p or $\bar{p}$	10 TeV (max) 10 TeV (max)	
DESY	HERA	e^- p	30 GeV on 820 GeV	1990
USSR				
IHEP	UNK	p $\bar{p}$	3 TeV 3 TeV	?
USA				
FNAL	Tevatron Collider	p $\bar{p}$	1 TeV on 1 TeV	1986
?	Superconducting, Super Collider, SSC or 'Desertron'	p p or $\bar{p}$	20 TeV on 20 TeV	?
SLAC	SLC	e^+ e^-	50 GeV 50 GeV	1986
Japan				
KEK	Tristan I	e^+ e^-	30 GeV on 30 GeV	1986
	Tristan II	e^- p	30 GeV on 300 GeV	1989?

CERN : Centre Européen pour la Recherche Nucléaire (Geneva)
DESY : Deutsches Elektronen-Synchrotron (Hamburg)
IHEP : Institute for High Energy Physics (Serpukov)
FNAL : Fermi National Accelerator Laboratory (Chicago)
SLAC : Stanford Linear Accelerator Center (Stanford)
KEK : Japanese National Laboratory of High Energy Physics (Tokyo)

Where do we go from here? We have been considering some of the lines of theoretical particle physics that are being actively pursued. The experimental programme is also very buoyant and, although there has been a sequence of remarkable discoveries in the last few years, there are plenty more prizes waiting behind the scenes. We expect further details concerning the top quark quite soon and hopefully a Higgs boson may appear. We may find SUSY particles and clear examples of proton decay may be seen. All of these discoveries could be crucial steps in our gradually improving understanding of the particles and their interactions.

A glance at Table 7.4 will show that particle physicists are optimistic about the progress that can be expected. A whole new round of accelerators (in the form of colliders) is planned and machines will soon be built all over the world which will take us into new energy regions and hopefully will give us not only some of the expected results, but also more surprises.

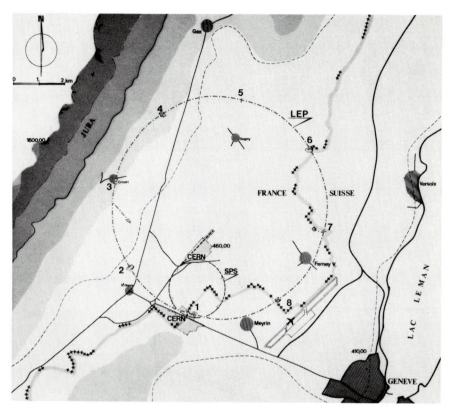

Figure 7.3. The LEP ring shown in relation to the Geneva area. This e^+-e^- collider will have a machine circumference of 27 kilometres.

APPENDIX A

physical constants, units and relativity

A.1. Physical constants

(All given to 4 significant figures)

Velocity of light	c	$2\cdot988\times10^8\,\mathrm{m\,s^{-1}}$
Magnitude of charge on electron	e	$4\cdot803\times10^{-10}$ e.s.u.
		$1\cdot602\times10^{-19}\,\mathrm{C}$
Planck's constant	$\hbar=h/2\pi$	$1\cdot055\times10^{-34}\,\mathrm{J\,s}$
		$6\cdot582\times10^{-22}\,\mathrm{MeV\,s}$
	$\hbar c$	$1\cdot973\times10^{-11}\,\mathrm{MeV\,cm}$
		$197\cdot3\,\mathrm{MeV\,fm}$
Mass of the electron	m_e	$9\cdot110\times10^{-28}\,\mathrm{g}$
		$0\cdot5110\,\mathrm{MeV}/c^2$
Mass of the proton	m_p	$1836\times m_e$
		$938\cdot3\,\mathrm{MeV}/c^2$
Bohr magneton	μ_B	$0\cdot5788\times10^{-14}\,\mathrm{MeV\,gauss^{-1}}$
Nuclear magneton	μ_N	$3\cdot152\times10^{-18}\,\mathrm{MeV\,gauss^{-1}}$
Avogadro's number	N_0	$6\cdot022\times10^{23}\,\mathrm{mol^{-1}}$
Fine structure constant	α	$1/137\cdot0$
Gravitational constant	G	$6\cdot672\times10^{-11}\,\mathrm{N\,m^2\,kg^{-2}}$

A.2. Units

(as used in particle physics)

Energy

electron volt $\qquad\qquad$ $1\,\mathrm{eV}=1\cdot602\times10^{-19}\,\mathrm{J}$
$1\,\mathrm{MeV}\ =\ 10^6\,\mathrm{eV}$
$1\,\mathrm{GeV}\ =\ 10^9\,\mathrm{eV}$
$1\,\mathrm{TeV}\ =\ 10^{12}\,\mathrm{eV}$

Length

femtometre $1\,\text{fm} = 10^{-15}\,\text{m}$
(often known as the 'fermi') $= 10^{-13}\,\text{cm}$

Cross-section

barn $1\,\text{b} = 10^{-28}\,\text{m}^2$
                                    $10^{-24}\,\text{cm}^2$

$1\,\text{mb} = 10^{-27}\,\text{cm}^2$

$1\,\mu\text{b} = 10^{-30}\,\text{cm}^2$
$1\,\text{nb} = 10^{-33}\,\text{cm}^2$

Mass

MeV/c^2 (If mass is m, rest energy is mc^2 MeV. Often mass is loosely quoted in 'MeV'.)

Momentum

MeV/c

Electric charge (and fine structure constant)

e in coulombs, then in SI, $\alpha = e^2/4\pi\,\epsilon_0\hbar c$
e in e.s.u., then in cgs, $\alpha = e^2/\hbar c$
e in Heaviside units, then $\alpha = e^2/4\pi\hbar c$
    $(\epsilon_0 = \mu_0 = 1)$
e in natural units, then $\alpha = e^2/4\pi$
    $(\hbar = c = 1)$
In all cases, α is dimensionless and $= 1/137\cdot 0$.

'Natural Units'

Natural units with $\hbar = c = 1$ are very often used in particle physics as h's and c's occur repeatedly. To get back to normal units we often make use of the relation $1\,\text{fm} \equiv (197\cdot 3\,\text{MeV}/c)^{-1}$.

A.3. Relativity — summary of some useful results

Further details may be found for example in H. Muirhead, *The Special Theory of Relativity*, Macmillan, London 1973.

Basics

Velocity of light in vacuo is the same in all inertial frames.

All laws of nature maintain their same form in all inertial frames (i.e. are 'covariant').

A point in space–time (x, y, z, t) is transformed from one inertial frame into another moving at uniform velocity, v, along its x-axis according to the Lorentz transformation:

$$\begin{aligned}
x' &= \gamma\,(x - vt) \\
y' &= y \\
z' &= z \\
t' &= \gamma\,(t - vx/c^2)
\end{aligned} \tag{A.3.1}$$

where $\gamma = (1 - \beta^2)^{-1/2}$ and $\beta = v/c$.

Four-vectors

The space–time coordinates form a 'four-vector', $\mathbf{x}$, with components $x_\mu\,(\mu = 1,2,3,4)$ such that $\mathbf{x} = (x, y, z, ict)$. The square of this four-vector is invariant and equals $(r^2 - c^2 t^2)$. Other four-vectors can be found (whose components, therefore, transform exactly like space–time components). For example, a momentum–energy four-vector can be defined as $\bar{\mathbf{p}} = (p_x, p_y, p_z, iE/c)$ and the transformation is

$$\begin{aligned}
p_x' &= (p_x - vE/c^2) \\
p_y' &= p_y \\
p_z' &= p_z \\
E' &= (E - vp_x)
\end{aligned} \tag{A.3.2}$$

To preserve conservation of energy and momentum, we must have

$$p = \gamma m_0 v \tag{A.3.3}$$
$$\text{and } E = \gamma m_0 c^2 \tag{A.3.4}$$

with the relativistic mass, m, given by

$$m = \gamma m_0 \tag{A.3.5}$$

We then have the relation

$$E^2 = p^2 c^2 + m_0^2 c^4 \tag{A.3.6}$$

and the invariant for the four-vector $\bar{\mathbf{p}}$ is $\bar{\mathbf{p}}.\bar{\mathbf{p}} = (p^2 - E^2/c^2)$ which is equal to $-m_0^2 c^2$.

Natural units

In natural units, with $\hbar = c = 1$, the most important of the above relations have the form:

$$
\begin{aligned}
\beta &= v \\
E &= \gamma m_0 \\
p &= \gamma \beta m \\
and \ E^2 &= p^2 + m^2
\end{aligned}
\qquad (A.3.7)
$$

Time dilation

An unstable particle moving with velocity v decays in a mean time, τ_0, in its own frame of reference. As observed from the stationary laboratory frame, a mean life τ is seen such that

$$
\tau = \gamma \tau_0 \qquad (A.3.8)
$$

Thus the observed lifetime is longer than the 'true' lifetime of the particle.

Available energy

If two particles with masses m_1 and m_2 collide, the energy available to produce a third particle is given by the total energy in the centre of mass (cms) frame, $\sqrt{s}$.

If the particle m_2 is at rest in the laboratory frame, and particle m_1 has energy E, then we can evaluate the available energy by using the invariance property of the four-vector; i.e.

$$
(\Sigma \, \mathbf{p})^2 - (\Sigma \, E)^2
$$

is the same in both the laboratory and cms frames, using natural units. Hence

$$
\begin{aligned}
p_1{}^2 - (E_1 + m_2)^2 &= -s \\
\text{or} \ s^{1/2} = m_1{}^2 + m_2{}^2 &+ 2m_2 E
\end{aligned}
\qquad (A.3.9)
$$

For energies such that E is much bigger than either masses m_1 or m_2, we have the approximation

$$
s^{1/2} \simeq (2m_2 E)^{1/2} \qquad (A.3.10)
$$

APPENDIX B

summary of non-relativistic quantum mechanics

This appendix gives a very brief review of some important results in non-relativistic quantum mechanics. A useful introduction to the subject can be found in, for example, R. M. Eisberg and R. Resnick, *Quantum Theory of Atoms, Molecules, Solids, Nuclei and Particles*, Wiley, New York, 1974.

B.1. Basics

Heisenberg uncertainty principle

Some pairs of observables are such that measurement of one disturbs the other, and the order matters if both measurements are made. In particular for

$$
\begin{aligned}
\text{position } x, \text{ momentum } p \quad & \Delta x \, . \, \Delta p \; \geqslant \hbar \\
\text{time } t, \text{ energy } E \quad & \Delta t \, . \, \Delta E \; \geqslant \hbar
\end{aligned}
\tag{B.1.1}
$$

where Δx is the uncertainty in the measurement of the variable x, etc.

Hypotheses of old theory

Planck's hypothesis: A photon of frequency ν has an energy E given by

$$E = h\nu \tag{B.1.2}$$

(From equation A.3.6, we have $E = pc$ and therefore $p = h\nu/c = h/\lambda$.)

de Broglie's hypothesis: A free particle with momentum p is associated with a wave of wavelength λ, such that

$$\lambda = h/p \tag{B.1.3}$$

If k is the wave number ($k = 2\pi/\lambda = 1/\lambda$) then

$$\mathbf{p} = \hbar \mathbf{k} \tag{B.1.4}$$

111

Wave-functions

The de Broglie wave for a beam of free particles with momentum p and energy E is represented by the wave-function

$$\Psi = \exp(i(\mathbf{p} \cdot \mathbf{r} - E \cdot t)/\hbar) \qquad (B.1.5)$$

where non-relativistically $E = p^2/2m$ for a particle of mass m.

In general we interpret $|\Psi|^2 \, dv = \Psi^* \Psi \, dv$ to be the probability of finding a particle in a volume dv at time t. Since Ψ is complex, we obtain a real probability by taking the absolute square of Ψ, where Ψ^* is the complex conjugate of Ψ.

B.2. The wave equation

The wave function Ψ for a single particle of mass m moving in a potential V satisfies the Schrödinger wave equation

$$\left[-\frac{\hbar^2}{2m} \nabla^2 + V \right] \Psi = i\hbar \frac{\partial \Psi}{\partial t} \qquad (B.2.1)$$

This involves partial differentiation with respect to time t and the spatial variables. The operator ∇^2 is in Cartesian coordinates a shorthand notation for

$$\frac{\partial^2}{\partial x^2} + \frac{\partial^2}{\partial y^2} + \frac{\partial^2}{\partial z^2} \qquad (B.2.2)$$

Standing wave solutions can be found from the time-independent Schrödinger equation

$$\left[-\frac{\hbar^2}{2m} \nabla^2 + V \right] \Psi_n = E_n \Psi_n \qquad (B.2.3)$$

with allowed energies E_n.

This equation can be solved with the relevant boundary conditions.

Examples

(*a*) A particle trapped in a rigid cubic box of side L:

$$E = \pi^2 \hbar^2 (n_x^2 + n_y^2 + n_z^2)/2mL^2 \qquad (B.2.4)$$

where n_x, n_y and n_z are positive integers (i.e. three 'quantum numbers' defined for this three-dimensional case).

(b) A particle bound in a coulomb potential: wave-functions of form

$$\Psi_{n,l,m} = N \cdot R_{n,l}(r) \cdot Y_{l,m}(\Theta, \Phi) \tag{B.2.5}$$

where R is the radial part of the wave-function and Y is the angular dependent part and N is a normalization factor. Again, n, l and m are the three quantum numbers with

$$
\begin{aligned}
n &= 1, 2, 3 \ldots \\
l &= 0, 1, 2 \ldots (n-1) \\
\text{and } m &= -l, \ldots -2, -1, 0, +1+2, \ldots +l
\end{aligned}
$$

The allowed energies are given by

$$E_n = -me^4/2\hbar^2 n^2 \tag{B.2.6}$$

(The above results are for a spinless particle)

(c) Particle in a harmonic oscillator potential:
Potential is of form $V(x) = \frac{1}{2}m\omega^2 x^2$.
Then allowed energies are given by

$$E = \hbar\omega(n + 1/2) \tag{B.2.7}$$

where n is a positive integer.

B.3. Operators and states

Every real observable A has an associated (Hermitian) operator $\hat{A}$ such that if

$$\hat{A}\psi_n = a_n\psi_n \tag{B.3.1}$$

then the a_n's are the allowed values of A (eigenvalues) and the ψ_n's are the allowed states (eigenstates) of the system. Thus, the results of a measurement on A must be one of the eigenvalues, a_n.

The wave-function can always be represented as a linear sum of the ψ_n eigenfunctions, i.e.

$$\Psi = \sum_n c_n \psi_n \tag{B.3.2}$$

where the c_n's are constant coefficients.
The ψ_n's form a 'complete orthonormal set', i.e.

$$\int \psi_n^* \psi_m dv = 1 \;\; ; \;\; n = m, \text{ normalized}$$
$$= 0 \;\; ; \;\; n \neq m, \text{orthogonal}$$

For components of momentum, **p**, and position, **r**, we choose

$$\hat{p}_x \rightarrow -i\hbar \frac{\partial}{\partial x} \quad \text{and} \quad \hat{x} \rightarrow x \tag{B.3.3}$$

with similar relations for the y and z components.
The total energy operator (the 'Hamiltonian') is

$$\hat{H} = i\hbar \frac{\partial}{\partial t} \tag{B.3.4}$$

The relations between the operators are the same as those between the classical observables. Hence

$$E = \text{kinetic energy} + \text{potential energy}$$
$$\text{or } E = p^2/2m + V(r)$$

Replacing the observables by their operators acting on the system represented by the wave-function Ψ, we have

$$i\hbar \frac{\partial \Psi}{\partial t} = (-\hbar^2 \nabla^2/2m + V(r))\Psi$$

which is the Schrödinger equation (B.2.1) again.

The operators representing the act of measurement do not necessarily commute (corresponding to the idea that the order of measurement matters). We define the 'commutator' of $\hat{A}$ and $\hat{B}$ as

$$[\hat{A},\hat{B}] = \hat{A}.\hat{B} - \hat{B}.\hat{A} \tag{B.3.5}$$

e.g., for momentum and position,

$$[\hat{p}_x, \hat{x}] = -i\hbar \tag{B.3.6}$$

which is seen to be non-zero and this is related to the uncertainty principle which acts for these observables.

B.4. Spin angular momentum

If the spin angular momentum is **S** and the orbital angular momentum is **L**, the total angular momentum is a vector sum

$$\mathbf{J} = \mathbf{L} + \mathbf{S} \tag{B.4.1}$$

The magnitude of **S** is $[s(s+1)]^{1/2}\hbar$ and

$$S_z = m_s \hbar \qquad \text{(B.4.2)}$$

where s is the integer (boson) or half-integer (fermion) total spin quantum number and m runs from $-s$ to $+s$ in integer steps.

The $s = 1/2$ and $s = 1$ cases are illustrated in Figure B.1. It is only possible to know **S** and one of the components, say S_z, simultaneously, the azimuthal angle being undetermined (i.e. it is possible to find wave-functions which are simultaneously eigenfunctions of the total and third-component spin angular momentum operators). It can also be shown that

$$[S^2, S_z] = 0 \qquad \text{(B.4.3)}$$

which is another way of expressing the same fact. On the other hand S_x and S_y do not commute with S_z so that they cannot be simultaneously measured.

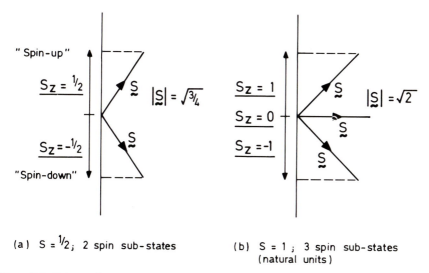

(a) $S = 1/2$; 2 spin sub-states

(b) $S = 1$; 3 spin sub-states (natural units)

Figure B.1. Spin angular momentum.

B.5. *Fermions and bosons*

Let ψ be the wave function which describes the state of a system of two identical particles, 1 and 2. Then $|\psi|^2$ will determine the probability of finding the two particles in some particular positions in space. If we now interchange the two particles, as they are identical we will have the same value of $|\psi|^2$ but the phase of ψ might have changed: $\psi \rightarrow e^{i\eta}\psi$. If we interchange the two particles again we must be back to the original situation so that $e^{2i\eta} = 1$ and hence $e^{i\eta} = +1$ or -1. For identical bosons we take the choice $\psi \rightarrow +\psi$ and for identical fermions we take $\psi \rightarrow -\psi$. Thus, under the exchange of two bosons we have a symmetric transformation but for the exchange of two fermions it is antisymmetric.

In general, there will be both space and spin components which are multiplicative in the wave-function. The spin component will be symmetric for two parallel spins and antisymmetric if the two spins are antiparallel. We therefore have the possibilities shown in Table B.1.

Table B.1. Particle exchange symmetries.

		With spins:	and	Space component:	Wave function
Fermions		Parallel	and	Antisymmetric	Antisymmetric
	or	Antiparallel	and	Symmetric	
Bosons		Antiparallel	and	Antisymmetric	Symmetric
	or	Parallel	and	Symmetric	

As the total wave-function of a pair of fermions must be maintained antisymmetric under interchange, it is impossible for them both to be in the identical quantum state which would necessarily give an overall symmetric wave-function. This is the basis of the Pauli exclusion principle.

FURTHER READING
a short selected list

Introductions to the theory of fundamental particles

The Nature of Matter, J. H. Mulvey (ed.) (Oxford: Oxford University Press, 1981)
Quarks, H. Fritzsch (London: Allen Lane, 1983)
The Particle Play, J. C. Polkinghorne (Oxford: Freeman, 1979)
The Cosmic Onion, F. E. Close (London: Heinemann, 1983)
Particles and Fields, W. Kaufman (ed.) (Scientific American Reprints, 1980)

Introductions to cosmology and particle physics

The First Three Minutes, S. Weinberg (London: Deutsch, 1977)
The State of the Universe, G. Bath (ed.) (Oxford: Oxford University Press, 1980)
The Nature of Matter, J. H. Mulvey (ed.) (Oxford: Oxford University Press, 1981)
The Cosmic Onion, F. E. Close (London: Heinemann, 1983)

More advanced theoretical books

Elementary Particles and Symmetries, L. Ryder (London: Gordon & Breach, 1975)
Symmetry Principles in Elementary Particle Physics, W. M. Gibson and B. R. Pollard (Cambridge: Cambridge University Press, 1976)
Introduction to High Energy Physics (2nd edn), D. H. Perkins (London: Addison Wesley, 1982)
Introduction to Quarks and Partons, F. E. Close (London: Academic Press, 1979)
Gauge Theories in Particle Physics, I. J. R. Aitchison and A. J. G. Hey (Bristol: Adam Hilger, 1982)

Accelerators and particle detectors

Particle Accelerators, M. S. Livingston and J. P. Blewett (McGraw Hill, 1962)

Europe's Giant Accelerator, M. Goldsmith and E. N. Shaw (London: Taylor & Francis, 1977)

Spark, Streamer, Proportional and Drift Chambers, P. Rice-Evans (The Richelieu Press, 1974)

The Theory and Practice of Scintillation Counting, J. B. Birks (Oxford: Pergamon Press, 1964)

'Particle detectors', C. W. Fabjan and H. G. Fischer, *Rep. Prog. Phys.*, **43**, 1003 (1980)

'Bubble chambers', F. W. Bullock, *Sci. Prog. Oxf.*, **58**, 301 (1970)

Introduction to High Energy Physics, 2nd edition, D. H. Perkins (London: Addison Wesley, 1982)

INDEX

3